D1507515

The Equation for Excellence

ROLAND MEDIA DISTRIBUTION

The Equation For Excellence

How to Make Your Child Excel at Math

Arvin Vohra

Published and distributed by
Roland Media Distribution
www.RMDGlobal.net

First Edition
Layout and Cover Design by Burning House Art & Design
Text set in Caslon

Library of Congress Control Number: 2007941462
ISBN: 978-0-9801446-0-4

*This book is dedicated to my mother, who taught me to excel,
and to my father, who taught me to think independently.*

CONTENTS

In this text, the pronoun "he" should be understood to mean "he or she."
The methods discussed apply to both boys and girls.

1: WHY STUDY MATH?

When children ask why they need to study math, the answer usually has something to do with either daily life or applications to science and technology. The problem with the first motivation is that it is an obvious and transparent lie. The second type of "motivation" tends to have the opposite of the intended effect.

The "daily life" explanation tells students that they will need math for their daily activities. For example, they will need to calculate the tip in a restaurant, or determine how much they should pay for their groceries. Most students are quick to point out that this problem can be solved by carrying around a calculator. And anyone who is worried about running out of batteries can carry around a spare set of batteries, or even two calculators. Even cell phones have built in calculators.

The arguments against the "daily life" explanation continue. In daily life, you never need to do more than add, subtract, multiply, or divide. Why learn trigonometry? Why study calculus? Why do anything beyond arithmetic? Even math-oriented jobs rarely require any really advanced math. When I worked as an actuary, the only math I used on the job was multiplication and the occasional exponent (the actuarial profession is one of the most math-oriented professions in the world.)

The other rationale for studying math focuses on science and technology. We need math to design space shuttles and satellites, to work in laboratories, and to build the newest computers. In one way this argument makes sense. Much of that work requires intensive use of advanced math. But very few people work in those areas. Those that work in those areas usually do so because of an internal passion, not because of any external motivation.

In fact, from the perspective of most students, there is very little external motivation to be a scientist. The strongest external motivators for most teenagers are money, fame, power, popularity, and attraction to the opposite sex. None of these powerfully motivate students to pursue careers in science. For every million dollars a scientist makes, the businessmen for whom he works make a billion. For every famous scientist, there are a thousand famous musicians and actors. The scientists who made the nuclear bomb were not the ones to use it; that power belonged to politicians. And in American culture, scientists have no more popularity or sex appeal than anyone else.

Thus, this argument not only fails to motivate students; it actually does the reverse. A student with no interest in being a scientist who hears the technology argument now thinks that advanced math is useful only for scientists. Thus, he does not need to learn it. If his goal is personal gain, his time is better spent doing almost anything else – studying politics, learning to play the guitar, working out, or thinking of ways to make himself rich. Math becomes just an annoying requirement.

So then why should a student learn math at all?

Kings used to play chess to learn military strategy. When I first heard this at age ten, the idea struck me as unbelievably stupid. In chess the bishop can move only diagonally. The knight can move in an L shape. A real soldier, on the other hand, can move in any direction. How would studying chess help in any real war?

I had, of course, completely missed the point. Strategy has nothing to do with L shapes or diagonals. A chess player learns to anticipate his opponent. He learns to look for strong positions, rather than short term gains. He learns to make intelligent sacrifices, and be wary of the strategic artifices of his opponent. He learns to predict his opponent's future responses to his actions, rather than focusing on the immediate gains. This mental discipline makes his mind sharper, and he becomes a much more capable strategist.

Similarly, math is important not because it teaches a student how to use trigonometry to measure the height of a building, but because it develops a student's ability to analyze and solve unfamiliar problems. Math develops concrete reasoning, spatial reasoning, and logical reasoning. Math does not just develop skills that can be applied to science and technology; when math is taught right, it develops the student's fundamental cognitive architecture, increasing his intelligence. The student will develop the logical reasoning skills that allow a lawyer to analyze a legal situation and to present a coherent and convincing argument. He will develop the ability, essential for any businessperson, to isolate the key components of a system. He will develop mental skills that can be used in any problem-solving situation. His mind will become faster, sharper, and more precise.

What lifting weights does for muscles, math does for the mind. In no sport will an athlete suddenly lie down on his back and lift a weight ten times. However, the vast majority of athletes do the press. Why? It makes them stronger, and thus prepares them for athletic endeavors in general.

When you teach a child math in the right way, you are giving him the gift of a sharper and more powerful intelligence. You are helping him actually develop his mind. You are making him smarter. You are giving him the ultimate ability to succeed in the world, and to build a happier life for himself. You are not just making him better at math; you are making him better at thinking.

This book will show you how to make any student excel at math, even a student who is extremely lazy or innately bad at math. You will learn how to motivate any student and what to teach. Whether you are great at math or barely able to do algebra, there are techniques in this book that you can use.

There are a few things you need to know before continuing. The first is that the methods in this book are designed to be effective. They are not designed to be easy, nor are they designed to be fun.

On the flip side, this book does not advocate a "beat your kids to make them strong" type of approach. I never yell at any student, and I obviously do not use any physical punishment. If you do your part right, you will never need to yell at a student to teach him math.

Similarly, the techniques here are not ones designed to cause antagonism. Many of my students spend a good portion of their tutoring sessions frustrated with a math problem, begging for an answer, or literally groaning. And yet the ones who complain the most are the ones who seem to appreciate my training the most. In fact, many of those students pay for part of their tutoring fees with money received from allowances, jobs, and internships, rather than switch to a more moderately priced tutoring service. Instead of spending that money on entertainment, they voluntarily spend it to learn math.

Why would a teenager actually spend his own money to learn math? Because at some level every person desires ability more than entertainment. Although we often believe the opposite, most teenagers would rather gain intelligence than momentary enjoyment. I may make my students struggle more than another educator would. But my methods bring out their very best, and they can see it.

About half of this book focuses on motivation. Right now, some of the finest minds in the country are using every advertising trick they know of to persuade your child to act in certain ways. Alcohol and tobacco companies spend millions of dollars per year on advertising, as do hundreds of junk food, clothing, and entertainment companies. Thus, in today's world, weak motivational methods simply cannot compete. Parents and educators who want to be effective must use motivational methods as powerful as those used by today's professional persuasion artists.

In fact, you will have to be even more persuasive. Unlike an advertiser promoting entertainment or recreation, to effectively teach your child math, you will have to persuade him to take the more difficult, yet ultimately more rewarding, path.

For example, many schools allow students to use calculators. As this book explains, chronic calculator use can dramatically weaken a child's math abilities. Thus, you may be the one persuading your child to not use his calculator, even though his teacher encourages calculator use.

While that task might seem impossible, the methods in this book will show you what to do. Once students understand the damage that calculator use causes, most of them voluntarily stop using calculators altogether. Several of my students have even taken the SAT, the most important test of their lives, without calculators. Almost all of them returned with perfect SAT math scores.

The fact that you will be working to fundamentally improve your child's life will make motivation a bit easier. Even when kids complain, they know what benefits them. And over time, as they see themselves becoming more intelligent and more successful, motivation will become easier.

The rest of this book explains what to teach, and how to teach it. It explores the primary effective math teaching methods, including the legendary Asian system and the methods that underlie the success of my company, Arvin Vohra Education.

2: THE ASIAN SYSTEM

The belief that Asians are good at math is held with good reason: even in America, students with Asian parents tend to significantly outperform every other ethnic group. For example, 2005 math proficiency testing showed that Asian students had higher math proficiency scores than White, Black, and Hispanic students at all age levels *(Source: Child Trends Databank)*.

This section examines the techniques used by Asian parents and educators. Of course, there are variations depending on the country of origin and the individual, but there are techniques and principles that are almost universal among Asian parents and educators.

The Asian system is built on memorization. At an early age, children are taught to memorize multiplication tables and the like. As they get older, they memorize formulas, and even memorize step by step ways to solve specific problem types.

The Asian system is radically different from current American methods, which emphasize understanding over memorization. Where American parents and math teachers focus on explaining why a technique works, the Asian educators simply require that the student memorize the technique, and be ready to use it.

One might expect that such a technique would create students who simply have formulas memorized and are unable to understand what they are doing. But the reality is just the opposite. Once students have the information memorized, the understanding seems to come naturally. On the other hand, systems that drop memorization and focus on only understanding seem to have the reverse effect. Students often end up confused – unable to understand the problem, or to solve it.

This is one of the strangest paradoxes in math education, one that I wrestled with extensively at the beginning of my career as an educator. Why does memorization work in math? Why does focusing exclusively on understanding fail? Isn't math about understanding? Shouldn't memorization be saved for history?

To unravel this mystery, we will undertake a journey that will help us understand some of the most important cognitive principles involved in math education.

COGNITIVE OVERLOAD

Memorize the following list of words:
Cow, dog, horse, tree, sea, frog

Not too hard, right? Now memorize this list:
Frog, moss, grass, house, cow, mouse, deer, phone, well, spoon, table.

The fact that the list is longer makes it much harder. There are various memorization techniques a person can use to memorize the list, but it is not nearly as easy to memorize as the first list.

Most people can hold about 7 pieces of information in their working memory at any given time (usually between 5 and 9 items, depending on their complexity. Working memory is used to remember information for a short period of time. Long-term memory is used to remember information for years.) The first list only had 6 items. The second list had 11 items. But it was more than twice as hard to memorize. Why? It had gone over the limit.

Now let's look at an example that is a bit closer to math.

Here is a rule: When you see a cow, hit it with a frog.

Easy to memorize. Easy to understand. You might even find yourself remembering this "formula" several weeks from now.

Here is another formula:

When you see a ztyq, hit it with a tfgh. (Note: A ztyq is just a cow missing a leg. A tfgh is a frog with more that seven spots on his back.)

If you focus, you will be able to memorize this formula and this explanation for a few minutes. But you will probably forget it by tomorrow. There are two reasons for this. First, there are more items of information. Secondly, picturing this requires a bit more work.

If you have ever struggled with math, the feeling you get from the above "formula" may be familiar. Now try this:

When you see a mtyq that is lacking a tfgh, hit it with a mrtg. (A mtyq is a ftzz or a qoiu. A qoiu is a frog without feet. A mtyq is half a dandelion. The definition of a tfgh is given above. An fter is a half of a rofr, which is a cow's left hoof.)

This formula is extremely difficult to follow. Few people even bother to read the formula the whole way through, and those that do forget it quickly.

What does this have to do with math? Look at the following math problem:

Paint costs $3 for enough paint for one square foot. Fred wants to paint a rectangular wall that is 4 yards wide by 5 yards long. How much will it cost to buy enough paint for 3 coats?

If you know that the area of a rectangle is length times width, and that a yard is 3 feet, this problem should not cause cognitive overload. But if you do not have the facts and formulas memorized, you end up with:

Paint costs $3 for enough paint for one square foot. Fred wants to paint a rectangular wall that is 4 yards wide by 5 yards long. How much will it cost to buy enough paint for 3 coats? (The area of a rectangle is length times width. A yard is three feet.).

It looks familiar, right? We have not even come close to cognitive overload, but there is more to juggle now. Because the student must juggle the information in the problem and unfamiliar formulas, he is not able to focus exclusively on solving the problem.

Now look at this problem:

Fred wants to paint a can red. The can is a cylinder with height 20 inches and radius 10 inches. He wants to cover the sides of the can with three coats of paint and the top with four coats of paint. Paint costs 10 cents for enough to cover a square inch. How much will it cost, in dollars, for enough paint to paint the can?

This problem is a bit more complicated, but it is still only a prealgebra problem. Now look at what a student who does not know the formulas must juggle:

Fred wants to paint a can red. The can is a cylinder with height 20 inches and diameter 10 inches. He wants to cover the sides of the can with three coats of paint and the top with four coats of paint. Paint costs 10 cents for enough to cover a square inch. How much will it cost, in dollars, for enough paint to paint the can. (The top and bottom are circles; the area of a circle is π*radius². The radius is half the diameter. The lateral surface area is the circumference times the height. The circumference is π*diameter.)

Of course, in a real problem, the relevant information would not be neatly written in parentheses after the problem. The student would have to look it up, or ask a parent or teacher. He would not only have to juggle the information, but also keep it all together while he got it from different sources. He would have very few cognitive resources available to analyze and solve the problem, because his mind would be too occupied keeping the formulas straight. He would have little chance of getting the problem right; on a test, he would just hope for partial credit.

This problem would be given in a prealgebra class, usually to seventh or eighth graders. And yet many high school seniors would struggle with this problem. In fact, many adults would struggle with this problem. But the problem is not actually difficult; it just tends to create cognitive overload.

When a student hits cognitive overload, the signs are usually easy to see. He becomes visibly frustrated. He may act out emotionally, by yelling, crying, or swearing. This is often viewed as a deep-seated behavioral problem, but it often is not. The student is faced with an impossible situation that may seem pointless. How would you react if you woke up tomorrow morning locked in a cage, for no apparent reason?

Other students may withdraw, seeming as if they are somewhere else. Their faces may stop showing any expression, and they may show little reaction to instructions or questions. Teachers and parents often mistakenly conclude that such students are stupid. In reality, they are withdrawing from an incomprehensible situation.

Some students may write something completely random on the page, or blurt out a formula that has nothing to do with the problem. For example, they might just say "quadratic formula?" or "Pythagorean theorem?" They might even just guess a random number. A student who does this does not believe that his random utterance is the correct answer. In desperation, he just says something, knowing full well it is wrong.

If a student has been struggling for a long time, you are dealing with an even bigger problem. Remember this?

When you see a cow, hit it with a frog.

You know what a cow is, and you know what a frog is. So it is easy for you to understand the above rule.

With the cylinder problem above, the really struggling student sees something more akin to this:

When you see a faquat, hit it with a potwu.

Why? He might not know what a cylinder is. The phrase "lateral surface area" might as well be written in Babylonian. Diameter? Radius? Because he cannot picture the problem properly, the steps he must take are a meaningless series of commands. If by some miracle he remembers them for a quiz, he is sure to forget them by the exam. He is dealing with foreign concepts that he can not picture.

To picture the problem, he must store the following information in working memory:

1. What a cylinder is
2. What a circumference is
3. The formula for the circumference
4. The formula for area of a circle
5. What a radius is
6. What a diameter is
   6a. The relationship between radius and diameter
7. The formula for the lateral surface area (which is really just the area of a rectangle)

The student has hit seven before even starting the problem. His working memory is full, and the problem has not even begun! He has nowhere to store the information for the problem (what the height is, what the cost is, etc.)..

Additionally, when a student's cognitive resources are being fully used, the student is less able to check for random errors. The rate at which he makes careless mistakes goes up dramatically. In fact, a high number of careless mistakes is one of the signs that tell me that a student's cognitive resources are being overstretched during the problem-solving process.

This is a prealgebra problem. The difficulty increases as the student enters algebra, or moves up to calculus.

HOW THE ASIAN SYSTEM ADDRESSES COGNITIVE OVERLOAD

We discussed how working memory can hold about 7 pieces of information at one time. But you know more than seven facts. You know more than 7000 facts, for that matter.

That information is stored in long-term memory. The Asian system helps students store information in their long-term memory in such a way that it is readily accessible. To be more precise, the Asian system forces students to store information in their long-term memory, and to have it ready for use.

The Asian system is fantastically effective, and extremely simple. As soon as the child is able to talk, math training begins. The child is constantly taught to memorize math facts and drilled daily on the facts. He is quizzed constantly on his multiplication tables. He is quizzed constantly on formulas (e.g. area of a triangle, circumference of a circle, quadratic formula, etc.).

He is repeatedly given specific problem types until he can do them in a few seconds. For example, he might be asked to find the surface area of a cylinder every day. After a few weeks, he can find the surface area of a cylinder with incredible speed. With the constant drilling, the information is always readily accessible and is stored in long-term memory.

The student does not need to have any amazing innate intelligence. His intelligence can be just average. For that matter, it can be below average.

The results speak for themselves, but let's look at how this student analyzes the above problem. Remember, the formulas are so ingrained into his mind that he barely needs to think about them to use them. He has done problems like this one so many times that the process has become virtually automatic.

Here is the problem mentioned in the last section:

Fred wants to paint a can red. The can is a cylinder with height 20 inches and radius 10 inches. He wants to cover the sides of the can with three coats of paint and the top with four coats of paint. Paint costs 10 cents for enough to cover a square inch. How much will it cost, in dollars, for enough paint to paint the can.

Here is the Asian student's way of thinking about it:

Find the top area and multiply by 4. Find the lateral surface area and multiply by 3. Add the two areas, and then multiply the sum by 10 to get number of cents. Divide by 100 to get the number of dollars.

The correct mathematical steps are:

$\pi^*(10)^2 = 100\pi$ is the top area. Multiply this by 4 to get 400π. The lateral surface area is $2\pi(10)(20) = 400\pi$.
Multiply this by 3 to get 1200π. Add those two numbers together to get $400\pi + 1200\pi = 1600\pi$. Multiply this number by 10 to get

the number of cents, which is 16000π cents. Divide this by 100 to get 160π dollars. Note that π equals approximately 3.14.

COGNITIVE OVERLOAD IN ARITHMETIC AND ALGEBRA

We have seen how cognitive overload can be an issue when solving word problems. What about regular arithmetic and algebra problems?

Look at this problem:

$$\begin{array}{r} 45 \\ \times\ 37 \\ \hline \end{array}$$

Most adults would find this problem fairly straightforward. But what if you had not memorized your multiplication tables? Then rather than starting out by doing 7x5 = 35, your first step would be to add 5+5+5+5+5+5+5, to get 35. You would then carry the 3, and do 7+7+7+7 (instead of 4*7) to get 28, and the process would continue like that. You might even forget what you were doing before you finished, and have to restart. In other words, you would reach cognitive overload.

The chance of making a careless mistake would be pretty high. You might even be tempted to do 45+45+45+45... (37 times) rather than the step by step multiplication.

How about a harder problem?

$$\begin{array}{r} 4563 \\ \times\ 7452 \\ \hline \end{array}$$

A bit tougher. But if you did not know the multiplication tables, it would be incredibly difficult.

Keep imagining that you did not know the multiplication tables. Do this:

$$1/7 \ + \ 5/42$$

It is getting harder, right? If you do not know multiplication tables, it is hard to get started with this one.

Now let's move to algebra:

$$3/a \ + \ 4/(a+3)$$

This problem has nothing to do with multiplication tables. But a person who had not memorized multiplication tables would never have really understood how to add fractions. That person would never be able to understand how to do this problem. He would probably memorize some way of doing the problem in the short term, and forget it by the time the exam comes around.

The above problem is a beginner level algebra problem. More advanced problems would require the student to solve similar problems as just one part of a multi-step problem. As the difficulty increases, the problems become impossible for the child to attempt at all. The child has "slipped through the cracks." There is no way for the child to do the problem without cognitive overload. To understand and solve the problem, he would have to hold years' worth of material in his working memory, which is impossible. Alternatively, he would have to actually learn several years of math before attempting the problem.

Because the Asian system focuses on long-term memorization of basic math facts, those trained with this system never face cognitive overload on these types of math problems. In fact, the Asian system takes it one step further. Not only does the system force students to memorize multiplication tables, etc., it constantly drills students on basic problem types. The student would not just find it easy to figure out how to do the above algebra problem. He would not need to "figure it out" in the first place. He would have practiced similar problems so many times that the process would be virtually automatic; it would be no more difficult than walking.

DISTANCING

Remember this?
"When you see a cow, hit it with a frog."

That is a lot easier to remember than
"When you see a terqp, hit it with a srato."

Why? The two statements are equally complex. However, the first statement means something to you, while the second statement means nothing. You cannot visualize it, or make sense of it beyond committing it to rote. You can memorize it, but you do not really know what you have memorized. Your mind distances itself from the information. You might memorize the information, but it will never be fully incorporated into your understanding.

Let's see how this applies to math. Student A is a strong math student. Student F is a weak math student. Both are given the following formula:

> *Area of a circle is π*radius squared, where π =3.14159…*

They are both given the following problem:

The radius of a circle is 6. Find the area, in terms of π.

Both students do the following:

Area=radius^{2*} π

Area = 6^{2*} π

Area = 36 π

Internally, however, they did something completely different. Student A (the strong student) pictured the problem. Before he began, he visualized the following:

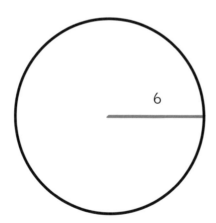

He knows what a radius is, and understands what he is doing as he calculates the area. If he is instead given a problem that says the diameter is 6 and asked to find the area, he will visualize this:

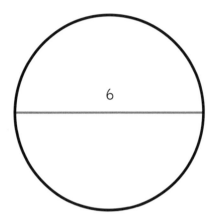

He will instantly see that the radius is 3, and will then solve the problem.

Student F (the weak student), did something very different. At some level, he thought "I don't know what a radius is, and I don't want to know. All I need to know is that if I am given a radius, I should multiply the radius by the radius, and then multiply by π."

I call this phenomenon "distancing", because the student distances himself from the concept. He keeps the information outside of his perception of the world. If the strong math student sees a pizza, he recognizes that it has a radius, and can picture the radius. The weak math student thinks of the word "radius" as descriptive only in his math class. He does not even consider that every circle he ever sees has a radius.

If Student F is given a problem that gives him the diameter instead of the radius, he is confused. Often, weak students end up just memorizing that the radius is the diameter divided by two, without visualizing the relationship. Not only does this further distance the student from the concept, it also gives him an extra item of information to memorize.

As math becomes increasingly complex, the student becomes increasingly distanced. He memorizes formulas by rote for each test, and comes close to failing each exam. The formulas mean nothing to him; he memorizes them by rote, just as you might remember

"When you see a terqp, hit it with a srato."

HOW THE ASIAN SYSTEM ADDRESSES DISTANCING

In the phenomenon known as distancing, the student uses temporary rote memorization to get by on math quizzes. The previous section explained why this is problematic.

At the same time, the Asian system is built on rote memorization. Students are often taught to memorize formulas well before they can understand them. A nine year old might memorize the quadratic formula well before he understands what the formula is used for. It would seem that this is a guaranteed way to create distancing.

However, it does the opposite. The information becomes fully integrated into the student's understanding.

Remember this?

"When you see a terqp, hit it with a srato."

It still means nothing, but it is starting to become more familiar.

How does the mind decide what information to incorporate and what information to "distance"? One consideration is the relevance of the information. Relevant information is easier to incorporate. For example, you might find the information in this book easier to incorporate than the information in a tax manual from the 1950s.

A second consideration is how interesting the information is. The pufferfish, which contains a highly potent neurotoxin, is considered a delicacy in Japan. Because of the potential dangers of eating pufferfish, it is the only delicacy forbidden to the Emperor of Japan. This fact is interesting, so it is easy to incorporate into permanent memory.

On the other hand, consider the following. The Black-Tailed Rattlesnake produces a highly toxic venom that is dangerous to man. This fact is less interesting, so it is easier to forget (even though it is simpler).

The next consideration is complexity. More complex information is more difficult to incorporate into memory. For example, it is easier to memorize "When you see a cow, hit it with a frog" than to memorize "When you see a cow, hit it with a frog, unless the cow is speckled, in which case hit it with a frog only if it is a weekend."

The final consideration is familiarity. Facts that are in some way familiar are easier to incorporate into memory than facts that are distanced. Take this for example:

"When you see a terqp, hit it with a srato."

It is still weird, but it is starting to sink in. If you saw this every day, and were constantly quizzed on it, you would have it memorized. And if one day you were given an explanation as to what it meant, it would be instantaneously incorporated into your understanding and long-term memory.

And that is how the Asian system prevents distancing. It prepares the mind to incorporate important information into its permanent understanding. By the time the child learns how to use the quadratic formula, it is so deeply ingrained into his memory that it becomes immediately incorporated into his mathematical understanding.

By constantly drilling information, the Asian system familiarizes the student with the information, preventing him from distancing himself from it. This prepares the student to instantaneously incorporate information into his permanent understanding as soon as he learns how to use the formula. Thus, paradoxically, by using rote memorization as a training tool, the Asian system prevents students from relying on short-term rote memorization, and instead makes them incorporate the information into their permanent understanding.

HIERARCHIZATION, AND ERRORS IN HIERARCHIZATION

One of the biggest differences between strong math students and weak math students is the way in which they hierarchize information (i.e., how they rank information in terms of its importance).

Strong math students generally hierarchize different concepts and formulas. For example, a strong math student recognizes the quadratic formula as an extremely important formula, and is able to easily recall it at any time. However, a formula of less importance, such as Descartes' rule of signs, gets less priority. The strong math student may be able to recall the rule after thinking about it for a few seconds, or he may even have to look it up. Similarly, the strong math student will instantly know how to factor

a^2-b^2 (the answer is $(a-b)(a+b)$.)

However, he may need to take some time to remember how to factor

a^3-b^3 (the answer is $(a-b)(a^2+ab+b^2)$.)

Even extremely strong math students, who are able to quickly do either problem, have the processes hierarchized. For example, the extremely strong math student will be able to do the first problem in about a tenth of a second, but may take up to two seconds to do the second problem. In other words, the second math problem will take 20 times as long as the first.

The weak math student, on the other hand, often hierarchizes the information incorrectly or not at all. In the first case, he gives undue weight to concepts of less importance, and insufficient weight to important concepts. In the latter case, he gives all concepts and formulas about equal weight. While he may actually remember unimportant concepts faster than strong math students, his skills with the important concepts are underdeveloped. While at the end of each year the strong math students remember about ten key concepts and are able to apply them flawlessly, weak math students have a vague and tenuous understanding of over a hundred math facts.

Students with ineffective hierarchization methods do very well on quizzes, poorly on tests, and horribly on exams. By the time they take the exam, their mind is so full of unimportant facts and formulas that they are completely overwhelmed.

These students often do extremely well in history classes, because of their capacity to remember large amounts of information for a few days or weeks. However, this very ability gets in their way in math classes. Because they can remember a huge amount of information for a period of weeks, they have less need to hierarchize information. Thus, they learn the information in an unhierarchized manner, giving the same weight to vital and unimportant information, and forget the important concepts along with the unimportant ones.

When taught concepts that are completely unimportant, strong math students will sometimes perform worse than weak math students. They instinctively recognize the information as irrelevant, and find it almost impossible to memorize the formulas.

HOW THE ASIAN SYSTEM HANDLES HIERARCHIZATION

American math classes and textbooks rarely employ effective hierarchization. Instead, most courses and books present essential information mixed in with a large amount of unimportant filler information. One week a student may study factoring, which is extremely important. The next week the class may focus on stem and leaf plots, which are relatively unimportant (a stem and leaf plot is a rudimentary way to organize statistical data.) The strong math students develop an intuitive ability to thoroughly understand the important information, and deemphasize the unimportant. Weak math students do not develop this ability, and thus struggle through math classes.

On the other hand, the Asian system does not rely on a student's innate ability to hierarchize information. Instead, the information is presented in a hierarchized manner.

As previously discussed, the Asian system drills students constantly. However, it does not drill students randomly. Students practice adding fractions. Students are quizzed on the quadratic formula. Students are quizzed on basic derivatives and integrals. Students are quizzed on the definition of sine, cosine, and tangent. Remember that they are not just quizzed right before a test on the subject. They are quizzed all the time. A ten year old student may be quizzed on the definition of sine. His first school test that asks the definition of sine may not be for another five years.

However, students are not quizzed on information of secondary importance. Students are not given stem and leaf plot practice problems, except right before a school test on the subject. Students are not quizzed on Descartes' Rule of Signs (a formula often taught as part of precalculus). It is not that these concepts have no importance; they just do not have the fundamental importance of skills like multiplying fractions or factoring.

The Asian system does not wait for students to hierarchize information. Instead, by strongly emphasizing important information, the Asian system ensures that students develop appropriate hierarchies.

Over time, the Asian method actually helps students develop their own hierarchization skills. Because important information is strongly emphasized, they begin to develop an intuitive understanding of what type of information is important in math. As they study additional topics in math, they learn to rely increasingly on their own ability to hierarchize information.

HOW THE ASIAN SYSTEM ENSURES UNDERSTANDING

We have seen how the Asian system enables students to solve problems efficiently. We have seen how the Asian system prevents distancing and ensures correct hierarchization. But what about understanding? How does the Asian system ensure that students actually understand what they are doing?

As you become more and more familiar with something, your mind becomes more comfortable exploring it. In math, students are more comfortable exploring familiar topics than unfamiliar ones. As they explore the topics consciously and unconsciously, their understanding of the topic automatically increases.

The Asian system, then, does not always teach students to understand the topic directly, at least initially. Rather than giving elaborate explanations, the Asian system focuses on making sure students are so familiar with the material that they are comfortable exploring it on their own. The Asian system does eventually give explanations, but they often come after the student has mastered the mechanics. For example, students are taught the mechanics of multiplying fractions first, and then given an explanation about why the method works.

By the time a student gets an explanation, he is already familiar with the mechanics. This allows him to dedicate his full cognitive resources towards understanding the problem, instead of having to split his attention between understanding the concepts and learning the mechanics.

DOES THE ASIAN SYSTEM TURN STUDENTS INTO DRONES?

In the opening chapter, I discussed why students should study math. I explained that math develops the mind, teaches students to analyze and solve new problems, etc. But on the surface, the Asian system seems to do none of them. It seems to turn students into unthinking automatons. Yes, they are able to do specific math problems quickly, but for most people math is not important for itself. It is only important because of the ways in which it develops the mind. The Asian system trains the mind, and drills the mind. But does it develop the mind? Does it increase intelligence? Or does it just turn students into drones who can quickly do specific types of problems?

The final and most important piece of the Asian system addresses this concern. Drilling is essential. Practice is essential. Memorization is essential. But it is not enough. The student also needs challenging problems.

Challenging problems are the ones that take anywhere from 20 minutes to a week to figure out. These are the problems that force students to put their knowledge together in new ways and to really stretch their mind to figure out new problems. This process makes students smarter, not just better at math.

It is easy to make up challenging problems for younger kids: just give them slightly more advanced problems. For example, if they already know how to multiply single digit numbers, give them a two digit multiplication problem and have them struggle with it. Whether or not they solve it does not really matter. As long as they really struggle, their minds will be developing.

If you are good at math, you will probably be able to make challenging problems for older kids as well. But if you are not, you can use other sources for math problems, such as SAT I and SAT II math practice tests. These can be found in practice books for these tests, which can be found in almost any bookstore. The problems towards the end of a section are usually the hardest. For example, if a math section on the SAT has 25 questions, questions 23, 24, and 25 will usually be the hardest.

USING THE ASIAN SYSTEM

Here are a few guidelines that can help you get your child started with the Asian system.

1. Set aside a specific time every day for math. The standard is 1 hour per day, every day, for math practice. This is in addition to any school math homework, and runs year round, including vacations.

2. Traditionally, in addition to the daily hour of math preparation, parents randomly quiz their children on math facts and math problems. This can take place on car rides, during meals, etc.

3. Materials: You can use math textbooks and math workbooks. If you have excellent math skills, I recommend using the University of Chicago School Mathematics Project materials, commonly known as "Chicago Math." It is an "expert system" in that you really have to understand math well to be able to understand and use the system effectively.

4. Fanatic dedication. All children (even Asian children) initially resist the Asian system. They will point out that their friends do not have to do the extra math work, and will do whatever they can do get out of it. Make sure they do the extra training. Your child may never enjoy it, but he will quickly see the benefits.

5. Constant Review: During the process, make sure to repeatedly revisit older topics.

6. Focus on the basics. The hour a day should be spent on major problem types, such as adding fractions, rather than on

minor problem types, like stem and leaf plots.

7. More is better. Two hours a day is better than one hour a day. Three is even better.

8. Repeatedly remind yout child that math develops the brain, and that by doing the extra math they are becoming smarter than their peers. Many of my math students voluntarily put in several hours of math training per week in addition to homework and tutoring sessions, because they know that the training is something that benefits them, not an obligation to someone else.

You should start using the Asian system today. If your child can talk, start the process. And it is never too late. You can start using the Asian system with a child who is 17, or with an adult for that matter. In fact, many of my older students voluntarily put themselves through the Asian system, with excellent results, as do some of my younger students. However, for most younger students, you will need to make them do it. They will not like it, but it will make them much more successful at math and life.

USING COGNITIVE OVERLOAD

The Asian system helps students avoid cognitive overload. Interestingly, in some cases you can actually use cognitive overload as a powerful cognitive incentive.

Suppose a child insists on multiplying by adding, rather than by using memorized multiplication tables. For example, the child will do 19*6 by adding up 6+6+6+6+6+6+6+6+6+6+6+6+6+6+6

+6+6+6. This method is slow, arduous, inefficient, and painful to watch. Often, it is almost impossible to convince the child to use memorized math facts.

Of course the child cannot do a problem like 43*96 without using memorized math facts. But many parents and teachers hesitate to give the student a problem like that, if the student is still using the aforementioned inefficient method. They feel that if they give such a student a problem like 43*96, the student will feel totally overwhelmed.

And they are right! The student will feel totally overwhelmed. And that unpleasant feeling gives the student a powerful incentive to switch to the more effective method.

What I usually do is give the student a mix of easy and hard problems. Hard problems give the student the incentive to learn the more effective method, which they can then practice on both the hard and the easy problems.

3: SELF-PERCEPTION
AND POLARIZATION

SITUATION 1:

You just started seventh grade, and it is the first math class of the year. The teacher explains something, and you notice that others in the class seem to understand the concepts faster than you do. You are able to solve the problems, but it takes you a bit longer than the rest. By the end of the class, you are starting to think that you just might be bad at math.

The next day confirms it, as the other students seem to race ahead of you. In reality, they are at best 5% faster, but from your perspective it seems like they are at least ten times better at math than you are. At this point (two days into the year), you are already pretty sure that you are bad at math. By the end of the week, you are absolutely certain. In fact, being bad at math soon becomes part of your identity. When the teacher gives a challenging problem, you do not really try that hard to solve it. That is just not who you are. You are the kid who struggles with math; what chance do you have of solving a hard problem?

You do your homework, but if you cannot get a problem, you are not too concerned. Athletes play sports. Rockstars sing. Your role in life is to get math problems wrong.

Every so often, you get some evidence that does not fit your theory. You get a 9/10 on a quiz that everyone else in the class fails. But you and the rest of the class are now so sure that you are bad at math that you decide that the quiz is flawed. Everyone laughs about it. How did the bad math student do better than everyone else on the quiz? Even your parents think it is kind of funny.

From time to time you get a math teacher that tells you that you can be an excellent math student, and that there is nothing wrong with your abilities. Sure, he is something of an authority. But you have years of experience, the opinions of other teachers, and the opinions of your friends on the opposite side. Obviously, you assume that the heretical math teacher is wrong and that every other person on Earth is right.

SITUATION 2:

You just started seventh grade and it is the first math class of the year. The teacher explains something to the class, but it is something that your mom taught you over the summer. You already know it, so you are the first one to understand the concept. Everyone around you notices, and the teacher marks you as one of the strong math students. In fact, by the end of the class, you are officially one of the excellent math students. That is how the teacher and every other student has started to see you, and it is how you are starting to see yourself. By the end of the week, it has become part of your identity.

From time to time, you come across a homework problem you cannot immediately figure out. But what if someone else in the class figures it out? Athletes play sports. Rockstars sing. Your role in life is to get more math problems right than any other student, and you will fill that role no matter what it takes. It may take you five hours, but you will figure the problem out.

Every so often, you get some evidence that does not fit. On one quiz, you get a 5/10, and everyone else gets at least an 8/10. The dumb kid gets a perfect score. Obviously, there was something wrong with the quiz. Everyone thinks it is funny, and the dumb kid announces to the world that he is now officially smarter than you. Much hilarity ensues, and even your parents think it is kind of funny. Your dad tells you not to make a habit of it, but it is more of a joke than a warning. Of course you will not make a habit of it. That is not who you are.

Two things happened in each situation. The first was that the child quickly developed a perception of his own ability by comparing himself to the rest of his class. Note that self-perception was determined by ranking, not by measuring. The struggling child did not measure how far behind the rest of the class he was; all that mattered was that he was behind. He was in last place. Whether the child finished ten seconds behind the rest or ten minutes, he was at the bottom.

Secondly, in each situation the class sorted itself into a hierarchy, and this hierarchy reinforced the child's self-perception. In any group, various hierarchies will be established. The tallest kid in a group is the tall kid, even if he is only an inch taller than the rest. The fattest kid is the fat kid, even if he is not really that fat. The slowest kid is the slow kid, even if he is only slightly slower than the rest. The smartest kid is the smart kid, even if he is only slightly smarter than the rest. Small differences become exaggerated, and the group becomes polarized.

The good news is, of course, that to make a student see himself as a smart kid, he does not have to be that far ahead of his peers. He only needs to be slightly ahead; once he and the rest of the class perceive him as a smart math student, self-perception and polarization will become the child's powerful allies.

The simplest way to take advantage of this is to use part of the summer vacation to cover the first few weeks of material. Find out what textbook the child will use in the coming year, and then go over the first few chapters. Your child will start the year slightly ahead of the pack, and will be much more likely to see himself as an excellent math student (as will his teacher and peers).

However, if everyone else in the class is doing the same, you will have to do more to keep ahead. Here is a secret: approximately 100% of Asian parents use this method. One of the biggest reasons that Asian American students do so well in math is that their parents make them spend part of their summer (usually an hour or two every day) doing math.

CHANGING SELF-PERCEPTION

Summer, winter, and spring vacations are the easiest times to change a student's self-perception, and his position in the academic hierarchy of his class. While his peers squander their vacations, your child can move forward. Thus, when school begins, he will start at the head of the curve.

Also, by making your child work during the vacations, you are making him see himself as the type of student that works through vacations. He accepts extra training as part of his life, and that extra training will carry him far.

BEST OF THE WORST OR WORST OF THE BEST

Sometimes, parents who are considering different schools ask whether it is better to put the child in a more competitive school in which he will struggle or in an easier school in which he would lead the pack.

First, those are never the only two options. By training extra, the student can lead the pack in the more competitive school.

But let's pretend that those were the only two options. The child can either be the best student in an easy school or a mediocre student in a difficult school. Which option is better?

As we discuss the importance of self-perception, it is important to remember that there is such a thing as reality. The student who leads the pack in an easy school will see himself as a strong math student, and will usually be motivated to study hard. But he will lack the rigorous training that the tougher school can offer. Simply put, he just will not learn as much. He will not become as smart. And when he competes in the outside world, he will not be as able to succeed.

The person who finishes last place in an Olympic race is still a world class athlete. A student who struggles in a competitive school is still miles ahead of a student who excels in an easy school. Sure, the student in the harder school will struggle. But when he competes

against those who lacked rigorous educations, he will easily surpass
them.

4: INCENTIVE AND STRUGGLE: The Art of Developing the Mind

When taught right, math builds the mind in the way that lifting weights builds the muscles. But not all methods of teaching math do this equally; in fact, some of the more recently adopted methods of teaching math actually do the reverse. Not only do these methods fail to build cognitive skills, but they actually cause skills that the student has already built to atrophy.

Three things cause cognitive skills to develop. The first is age. Even with the worst education available, human biology will make a sixteen year old more intelligent than a two year old.

The second thing that causes cognitive skills to develop is exposure. Children who are exposed to interesting ideas and problem types can freely stretch their minds and explore new modes of thought. As a simple example, a child who plays with a Rubik's cube may develop a surer sense of three-dimensional reasoning.

The third consideration is incentive. A child who plays with a Rubik's cube may develop the foundations for strong spatial reasoning; however, without a strong incentive, he may never push his mind to the limit. If he cannot figure out how to solve the Rubik's cube, he will probably just give up.

For the mind to have the incentive to develop, two things are necessary. First, it must encounter a problem that it is unable to do; the process of figuring out how to solve this initially unsolvable problem causes the mind to develop. If a student is only given problems at his current ability level, what incentive does the mind have to improve? Just as lifting a half-pound weight will not make a person physically stronger, doing an easy math problem will not make a person mentally stronger.

Parents and teachers of gifted students often overlook this, and just allow them to work at a comfortable pace. The result is that the gifted students never get the opportunity to realize their full potential. Like natural athletes who never train hard, they end up squandering their innate talents.

No matter how smart a student is, he must be given some challenging math problems that he is initially unable to do. If he can solve the problem in five minutes, it is not a real challenging problem. An appropriate challenging problem should take anywhere from twenty minutes to a week to solve.

Once you have a sufficiently challenging problem, the next thing you need is incentive. If the student has no incentive to figure out a difficult problem, he will simply walk away. However, by understanding what motivates your child, you can design the right kinds of incentives. The following list includes some common motivators:

1. Desire to impress. If your child wants to impress you, and is willing to struggle through a difficult problem to do so, all you have to do is give him verbal approval when he solves a challenging problem.

2. Desire for self improvement. At some level, everyone wants to be better at everything. Everyone would like to be smarter and stronger. Sometimes this desire can get buried under other desires, but it never disappears. Tapping into this desire is the strongest method that I know of, and I favor this method above all others.

3. Material incentives. Parents may try to use money or toys as an incentive to work hard (for example, a student may be given $20 or a video game for every A he earns.) As discussed later in this book, this method is almost always completely ineffective.

4. Laziness. Using laziness is another one of my favorite ways to motivate students. Everyone likes to avoid work. If figuring out one particularly difficult problem will allow the student to avoid twenty hours of math work, he will put his full abilities into solving the problem.

Many parents unthinkingly do the reverse. The better the student does, the more work he ends up doing. Every correctly solved problem is rewarded with a more difficult problem. Students who are both clever and lazy often figure this out, and deliberately get problems wrong in order to avoid doing extra work. This phenomenon and the way to address it are discussed in depth later in Chapter 12.

Once you have a challenging problem and have given the student a strong incentive, all you have to do is sit back and watch him struggle. Be patient. Do not explain how to do the problem or give excessive hints. Stay completely calm, and wait for him to figure out how to solve the problem. At most, give one hint every five minutes.

Every second your child struggles with the problem, he is

becoming smarter. His cognitive abilities are increasing bit by bit. If you can consistently create struggle, you will slowly but surely improve your child's reasoning skills.

When I watch a child struggle with a math problem, I imagine a meter at a gas station, with the numbers rapidly moving up. Every split second that goes by, the number increases. I imagine that this number corresponds to the child's intelligence. Every second I let the child struggle, his intelligence is increasing.

Give the student time with each problem. Struggling with one problem for several hours benefits the student much more than sailing through ten problems in ten minutes.

As you watch your child struggle, remember that the cognitive skills developed are not just good for math. They are the skills that allow a person to analyze and approach any kind of situation.

THE PATH OF LEAST RESISTANCE

Suppose you want to teach your child how to add fractions. Let's look at a few hypothetical situations.

1. You teach the child to add fractions by hand, and do not teach the child how to add fractions with a calculator. For all future problems, he must add fractions by hand, and is not allowed to use a calculator.

2. You teach the child to add fractions by hand first. Once he is able to add fractions by hand, you then teach him how to add fractions with a calculator. For all future problems he has the choice of adding fractions by hand or using a calculator.

3. You teach the child to add fractions by hand and how to add fractions with a calculator at the same time. For all future problems he has the choice of adding fractions by hand or using a calculator.

4. You teach the child to add fractions with a calculator only. You do not teach him how to add fractions by hand. For all future problems, he must use a calculator.

Which is the best option? At first, option 2 looks the best. In this situation, the student learns to add fractions by hand and also becomes proficient at adding fractions with a calculator. This option seems to combine the best of both worlds.

Most people who choose option 2 will say that option 1 is the second best, option 3 is the third best, and that option 4 is outrageously irresponsible. However, for all practical purposes, options 2, 3, and 4 are identical.

Let's look at option 4 first. The child was never taught to add fractions by hand, and he has no incentive to add fractions by hand, since he can always just use a calculator. Thus, he will never learn to add fractions by hand.

In option 3, the child can always choose either to add fractions by hand or with a calculator. Both methods will give him the answer, but using the calculator is easier. Thus, the student will probably always use the calculator. Even if he initially learns to add fractions by hand, this skill will atrophy through chronic calculator use.

Finally, let's take a look at option 2. At first, the student is able to add fractions by hand. However, he is then always given the option to add fractions with a calculator. He will most likely always use the calculator, and never add fractions by hand. This will allow his skill with adding fractions to atrophy. After several years, he may entirely lose the ability to add fractions by hand.

I have worked with hundreds of high-school students from the best public and private schools in the Washington, DC area who can no longer add fractions by hand. They were able to manually add fractions at one time, but after 5 years of chronic calculator use, the skill has completely atrophied. Since students who cannot add fractions by hand cannot do several types of advanced problems (adding rational expressions, partial fraction decomposition, etc.), they eventually find themselves completely stuck and frustrated.

This principle does not apply only to fractions. In most situations, the mind takes the path of least resistance. It learns the easiest available method to solve a specific problem type, and forgets the rest. The other skills quickly atrophy.

It is important to note that students do not want their skills to atrophy. Similarly, an athlete who is bedridden for several months does not want his muscles to atrophy. It just happens. The human mind and body do not waste resources maintaining or developing unnecessary abilities.

LABOR-SAVING PATTERNS

When you do math by hand, your mind searches for patterns in order to simplify your future work. For example, you probably know that to multiply a number by 10, you can just add a zero to the end of the number (e.g. 37 * 10 = 370).

Some people incorrectly believe that students who are exposed to enough calculations will automatically begin to detect these kinds of patterns. Supposedly, even if a student does math on a calculator, he will begin to see these patterns and incorporate them into his thinking.

This view misses the obvious. Most people only look for these kinds of patterns when there is an actual incentive to do so. For example, we look for these patterns to avoid extra calculations. But if there is no incentive to do so, why would the mind waste energy and resources looking for patterns? Unless the student has some powerful and innate interest in numbers, he will never discover the patterns.

Searching for these patterns is in itself an excellent mental exercise. Equally importantly, these patterns allow students to quickly solve more difficult problems. A student who is using labor-saving patterns may solve a difficult SAT problem in 3 minutes. A student who is not using these patterns may eventually solve the same problem, but it may take him several hours to do so.

"I DON'T KNOW"

I have had the following conversation with dozens of students.

Student: I don't know what to do. I don't get it.
Arvin Vohra: You know how to do this problem.
Student: No, I don't
AV: Yes you do.
Student: Arvin, seriously, I don't know what to do.
AV: Yes you do.
Student: I have no idea. Just tell me.

AV: Nope.

Student: Is it…I have no idea. I forget.

AV: No you don't.

Student: Please?

AV: Nope.

Student: Can you show me an example?

AV: Nope.

Student: Can I have a hint?

AV: Nope

Student: Do I do this? [begins the problem correctly]

AV: Yes.

When a tutor, teacher, or parent hears the words, "I don't know," or "I don't get it," the first instinct is to explain how to do the problem. One percent of the time, that is exactly what should be done. The other 99% of the time, it only hurts the student.

When faced with a problem he does not immediately know how to solve, a student has two basic options. The first, and more difficult option, is to actually figure out how to do the problem. This option usually means a few minutes of struggle and maybe some frustration. Through this process, his mind is analyzing the problem from every direction, seeking some connection or opening that will allow him to solve it. These periods of intense mental strain build and sharpen his intelligence.

The other option, is to say "I don't know," and get an explanation from a parent or teacher. While he may learn how to solve that specific problem, he has lost the chance to develop his ability to analyze and solve unfamiliar problems. In other words, he has lost the chance to increase his intelligence.

As discussed earlier, the mind takes the path of least resistance. Most students in that circumstance will take the easier, and ultimately less beneficial, path.

But what if the student knew that no one would tell him how to do the problem? What if he knew that his only choice was to figure it out on his own? Then he would have no choice but to struggle, and thereby develop his mind.

Thus, when a child asks for an explanation for a math problem, 99% of the time you help him more by giving no explanation. By repeatedly forcing him to figure out problems on his own, you will gradually build his fundamental math reasoning skills.

This can require a lot of patience and discipline on your part. When you see a child struggling with a problem, the desire to just explain how to do the problem can be almost overwhelming. You will want to explain it to him, to make him see the answer. Watching him struggle may initially be almost as frustrating for you as it will be for your child. Just wait. Do not show any signs of losing patience. Make sure that he knows that you have no intention of caving in and giving an explanation. There is absolutely nothing wrong with spending twenty minutes on a single math problem, especially since those twenty minutes will develop his mind far more than twenty hours of hearing explanations.

Sometimes, you can use even more powerful incentives to make the process more efficient. One of my favorite drills is the Hydra, which is also discussed later in the book. Every time the student forgets how to do a problem, he gets two additional similar problems. If he gets one of those wrong, he gets two more, etc. Once he has eight extra problems, I give him a hint. Once he gets sixteen extra problems, he gets a second hint. This gives his unconscious mind a powerful incentive to learn how to do the problem. (The Hydra is based on the mythical Greek monster with many heads. If one of the heads was cut off, two more grew back in its place.)

The Hydra is an extremely powerful motivator. When a student who asks for a hint gets two additional problems instead, he immediately puts his full mental resources into solving the problem.

However, for longer, more difficult problems, the Hydra does not make sense, and it is best to just let the child struggle. Sure, you can use the Hydra to ensure that the student learns the laws of exponents. But it rarely makes sense to use the Hydra when dealing with long geometry proofs. For such complicated problems, just let the student struggle through the problem.

EASY COME, EASY GO

When I first started tutoring, I would give an explanation any time I heard "I don't know." I quickly noticed that I was explaining the same things to the same students again and again. I was baffled. My instincts told me that these students were smart. Then why were they completely unable to remember anything for even a week?

The reason was, of course, that I had not given the student any real incentive to remember the information. The students could get an explanation any time they wanted. Why should they remember how to do the problem themselves?

Students who figure out how to do a problem on their own almost never forget how to do it. They will still be able to do the problem years later. On the other hand, a student who is given the explanation often forgets it within a day, especially if he knows that he can get the explanation again any time he wants. Since they have little incentive to incorporate the information into their permanent understanding, many students who receive explanations for every problem do horrendously on exams and standardized tests.

Sometimes, I will give a student a hint. In the extremely rare situations in which I do tell a student how to do the problem, I first make him struggle for several minutes. I may give him several extra problems before even the slightest hint. Then, when the explanation is given, it comes as such a relief that he remembers it forever. Additionally, if the student forgets, I do not explain the problem the second time, but instead make the student figure it out. This ensures that the student develops the cognitive skills required to understand how to solve the problem.

I give a full explanation less than one percent of the time. The first time a student sees a specific problem type, I might tell him how to do it. For example, if a student has never multiplied fractions before, I may show him how (after making him struggle for a few minutes. Most students are actually able to figure out what to do on their own, though you will have to let them know when they get the answer right.)

Note that when a child says "I don't know," he is telling the truth. He probably does not know how to do the problem right away. It will probably take him several minutes to figure it out. Your job is to make sure that he has the incentive to figure out the problem on his own. As long as he has the incentive to put his full mental abilities to the task, he will be developing his ability and intelligence.

However, as the saying goes, a reputation takes a lifetime to build and a second to lose. Cave in to even one "I don't know," and the child will know that he can extract help from you. He will then have the incentive to try to turn this into a regular pattern. Effort that could have been spent on developing the child's mind is now wasted as he tries to get the answer in another way.

If you never give an explanation, and always make the student figure the problem out for himself, you will be doing the right thing 99% of the time. You will be helping him develop his cognitive skills, along with his independence, self respect, and confidence.

THE OTHER EXTREME

Not giving the answer does not mean letting the child do whatever he wants. If you just refuse to give an answer, and then leave the room, the student may give up and just find a way to entertain himself for an hour. It is still important to make sure that he is working on the problem.

Finally, it is important to check the work. Remember, the mind takes the path of least resistance. If a student knows that the work will not be checked thoroughly, he may be strongly tempted to just write some random math steps down on the page. Many math text books contain answers to odd-numbered problems, so it is often a good idea to have the child do only the even-numbered problems. Check to make sure each step is correct, rather than only checking the final answer. Once you get the hang of it, it will not take very long to check each problem.

Do not rely on schoolteachers to check homework. In many schools, teachers do not check the homework for accuracy. That means that a student can get every homework question wrong, or just write some random numbers on a page, and still get full credit. Unless your child is lucky enough to have an unusually rigorous teacher, it will probably be up to you to make sure that your child is getting the right education.

5: HOMEWORK:
Daily Motivation

A toddler stumbles and mildly bumps his knee. He immediately looks to his parent. If the parent responds with extreme shock and concern, the child starts crying. On the other hand, if the parent simply reacts as if the fall was no big deal, the child brushes himself off and keeps playing.

The child is new to the world. He does not really know what reactions are socially expected, so he looks to the authority figure to determine the correct reaction to his current predicament.

If a child does not regularly finish his homework, there is only one reason: his parents do not really, fully, deep down think that skipping homework is a big deal. Of course, consciously, the parents think it is a big deal. They know that the child should do his homework every day. They know that the homework will make the child smarter and will lead to more opportunity. But some tiny, unconscious part is still willing to let homework slide.

Here is what I mean. Make a mental list of situations in which it would be okay to cut off your child's finger. It is probably a pretty short list, and might look like this:

1. If his life absolutely depends on it.

Now make a list in which it would be okay to allow your child to skip his homework. If your child always does his homework, the list is probably not much longer. In fact, the list might be exactly the same.

If your child regularly skips homework assignments, your list is probably noticeably longer. On the surface, this seems to make sense. After all, cutting off a finger is a major and traumatic surgery. But skipping homework – well that just does not seem like as big a deal.

But let's look at things objectively. A child missing a finger (or an arm, for that matter), still has a shot at attending a competitive college. He will still be able to compete against Students from all over the world in today's and tomorrow's global economies. He will still have a high chance of having a successful life.

However, a student who regularly skips homework is not in the same position. A student who falls behind in school does not have access to the same opportunities as one who excels. Thus, in a practical sense, skipping homework regularly can actually be much more damaging than losing a finger.

Once you fully accept the absolute necessity of homework, your actions will flow naturally. What would you do to stop your child from cutting off his hand? Whatever it took, right? Approach homework with the same perspective. Do whatever it takes to make sure your child does all of his homework every single day.

If you really, fundamentally understand that homework must be done, no matter what, the child will sense that, and will respond accordingly. If you do not, the child will sense that as well, and he may regularly skip homework.

SELF-PERCEPTION AND HABIT ARE MUCH MORE IMPORTANT THAN INCENTIVES

If you ask the parents of an outstanding math student what kind of rewards they use to motivate their child to do his homework, for a split second they will look at you as if you asked what kind of reward their child gets for not wearing diapers anymore. They will then quickly recover and come up with some polite answer that sounds good. But that initial look tells you everything you need to know.

Parents of outstanding math students only use incentives when the child is very young, and then only for a short period of time. The approach is almost like toilet training. When a child is first toilet trained, parents may give great approval, and even some small reward, to the child when he correctly uses the toilet. But very few parents cheer a 15 year old of normal intelligence when he manages to use a toilet correctly. By that age it is just taken for granted. Similarly, no approval is given when a child does his homework every day. It is just taken for granted.

So how do you get there? How do you make homework a permanent part of a child's daily life?

First, you have to be consistent. If something is a big deal, then it is always a big deal. If something is almost always a big deal, but occasionally not a big deal, then it is not a big deal. If your child cut his thumb off intentionally, that would be a big deal. That would be a big deal any day, no matter what. Give schoolwork the same level of importance, and watch your child excel.

Second, you have to make it possible for the child to do his homework. First, that means you need to give the child a table, good lighting, and a reasonably low noise level. The table must be the kind at which you can sit and work, such as a desk or dining room table. A coffee table or counter is not good enough.

The noise level must be reasonably low. Do not blast the TV or radio in the room in which the student is working. If the TV and table are in the same room, move one or the other. You do not need absolute silence, but it can be hard to work while people are watching TV, listening to music, and having loud conversations in the same room.

There must be enough light to read. If the room is dim, buy a lamp. I know this section is starting to sound obvious, but I have been in dozens of homes in which there was literally no comfortable way for the child to do his homework.

Third, if you want to make something into a habit, you have to make sure it happens at a regular time. When do you brush your teeth? Just before bed, and when you wake up. It is unlikely that you brush your teeth sometimes at 5 p.m., sometimes at 11 p.m., sometimes at 7 p.m., etc. Similarly, you must make sure the child does his homework at a regular time. If you tell the child to do his homework sometime between 3 and 11 p.m., it might not get done. If you make him do it between 5 and 7 p.m. every day, it will always get done. Make sure the child has a consistent time to do his homework, and make sure that it always gets done, without exception. Within a few months, it will become a permanent habit.

More importantly, it will become part of his self-perception. He will begin to see himself as one of the kids who does his homework, rather than as one of the kids who do not. As you learned in the previous chapter, self-perception is one of the most powerful motivators available.

THE BEST KIND OF "PUNISHMENT"

The most common types of punishments are either harmful (e.g. physical punishment) or neutral (making a child sit in a corner). Although these punishments have been used historically, there are better alternatives.

Athletic coaches often need to punish players for minor infractions (tardiness, unwashed uniforms, etc.) Interestingly, these "punishments" are almost always beneficial. For example, a player who has to do fifty extra pushups builds his strength. A team that has to run an extra mile builds endurance.

You can do the same with math. If a student forgets how to add fractions, make him do 100 extra fraction addition problems. In addition to motivating him to remember, you are helping him build the necessary skills. If a student forgets the quadratic formula, make him write it down 10 times.

Beneficial "punishments" are easy to carry out. You will always feel good about making your child do extra math problems. At the same time, your child will see that you have his best interests at heart. You will gain both his trust and respect.

CONCENTRATION CAN BE A LEARNED SKILL

If a child says he cannot swim, he is probably telling the truth. He cannot. He does not know how to swim. He knows the general idea of what swimming is, but he himself cannot do it.

However, that does not necessarily mean that the child has some physical defect. It does not mean he needs to walk around all day with a life vest on, just in case he falls into some water. It just means he needs swimming lessons, and then a lot of swimming practice.

Similarly, if a child says he cannot concentrate for two hours, he is telling the truth. But that does not necessarily mean that he has a mental defect. It just means that he needs to learn how to concentrate, and then practice concentrating.

My first student with severe Attention Deficit Hyperactivity Disorder (ADHD) had a high IQ and a total inability to concentrate. Not surprisingly, he was doing horribly on the SAT, which requires students to concentrate for several hours.

The solution was to build his concentration skills. I first had him do math for fifteen minutes without stopping. Then we increased the time to half an hour without stopping. Then we increased the time to one hour. Then we increased it to two hours. By the time he took the test, he was comfortable doing math for three hours without stopping. He ended up with a perfect SAT Math score.

Building up concentration is like learning to walk. It is a slow process, but it is a process that almost everyone can go through.

If your child has difficulty concentrating, help him develop the skill. Do not wait for him to sink to the bottom of the class. Do not treat him as if he is helpless, unless you want him to think that helplessness is the best way to approach life.

So what should you do? Treat concentration like any other skill, and help him gradually learn how to do it. For the first week, have him concentrate for 15 minutes at a time. Then increase the time to 30 minutes. Then increase it to an hour. Then increase it to an hour and a half. After two months, he should be able to concentrate for 3 hours.

From this experience he will gain far more than just the ability to concentrate. You will have taught him to take control of his life, and to refuse to accept perceived limitations. You will have taught him to overcome challenges and to do what is necessary to succeed.

The worst thing you can do when your child has difficulty concentrating is to exacerbate the problem by providing him with distractions while he works. Do not let him watch TV when he is doing his homework. Do not let him listen to the radio or play on the computer while he studies. If your child has difficulty concentrating, help him develop ability to concentrate; do not let difficulty in concentrating become a permanent part of his life.

If your child has had difficulty concentrating for several years, do not give up. It is never too late, as long as you are willing to do what is necessary. The student with ADHD mentioned earlier this section was 17 when I started working with him. By the time he was 18 he had a perfect SAT math score.

6: LUXURY AND LONG-TERM MOTIVATION

We learned in the last chapter that motivating a child to do his daily homework has more to do with self-perception than with incentives. However, to really bring out the best in a child, you will need to provide the right type of long-term incentives, in addition to helping him develop the right self-perception.

This is especially important for brilliant but lazy children who think outside the box. Because these students tend to look at situations objectively, self-perception does not have the same momentum with them as it does as with others. Thus, these students need the right objective, long-term incentives to bring out their best.

Why might a student not do his homework? Sometimes students look at the short-term gains from not doing their homework (immediate fun) and ignore the long-term benefits of strong cognitive skills. For other students, particularly for students with learning differences such as dyslexia and students with delayed intellectual development, the homework is such a painful process that they avoid doing it at all costs.

But in one of the most common types of situations, neither applies. The main details are usually the same. The child is raised in an upper middle class family, in a wealthy neighborhood. His parents are successful professionals, often doctors or lawyers, and he has seen the positive results of an excellent education. The large house in which he lives, the car in which he rides, the luxuries he enjoys resulted from educational success. If anyone on Earth should be convinced that education is important, it is this child.

As one looks more closely at the situation, it makes even less sense. The child is unusually bright, and his memory and processing speed are excellent. Even his attention span may be above average. This is not a case in which the child is overwhelmed by the homework, or unable to do it. Nor is the child too young to understand the consequences; typically the child is between 13 and 15. It seems as if the child just does not want to do any homework, so he does not do it. The parents (and teachers) are often tearing their hair in frustration, as increasingly elaborate threats and bribes fail to motivate the student.

Why would an intelligent, able student who obviously knows the value of an education not do homework? Why would he not bother to study for a test? Is he just short sighted? Does he just not see the big picture?

Surprisingly enough, the answer is often just the opposite! The child is not really thinking only about the short-term; in his own way, he is thinking about the long-term.

Let's look at the situation from the child's perspective. His material needs are taken care of. In fact, the child's parents have the resources to support the child for life, if necessary. The child thinks it is unlikely that he will ever make enough money to significantly improve his material standard of living, and he is probably right. Whether he works diligently or never opens his textbook, there will probably be a minimal impact on his future lifestyle. He could only enjoy a significantly higher standard of living if he becomes a billionaire, and the child recognizes that this is unlikely. He has no real long-term incentive to work hard, so why should he?

Parents will often try to use various incentives to encourage the student to put in more effort. They might give the child presents or money for a good grade, or they might yell at him for a bad one. However, this method almost never works.

Why do these incentives fail? Suppose you had a job you hated. In this job, your salary is fixed, and you cannot be fired. Your choice is to work really hard for eight hours, or do no work at all and just get yelled at for twenty minutes every day. If you worked really hard all year, you might get an extra $100 bonus. Which option would you choose?

The child is facing a similar choice: 3 hours of work, or 15 minutes of being yelled at. The overall long-term effect is basically the same.

In the job example, if the punishment was 8 hours of electroshock for not working and $200,000 a day for working, you would probably work. Similarly, if the child was rewarded with a Ferrari for every good quiz grade, he would probably work hard. But in this chapter, we will see a more elegant system, one that will address all of the underlying problems.

ADDRESSING THE PROBLEM OF ASSURED LUXURY

As we have seen, the child is not short sighted. In fact he is thinking more about the long-term than his parents. While his parents are worried about his current test grades, he has already solved the problem of his future livelihood. He has figured out a way to not work and to have all of his material needs taken care of. He is free to do whatever he wants. He has outsmarted the system.

That means we need to change the system itself. Specifically, we need to change the long-term outcome. We need to make sure that he knows that his life will be much better if he gets a good job or starts a business than if he lives off his parents.

The basis of the child's current decision is assured luxury; he knows that he will be able to enjoy the material aspects of life.

How can a parent change this? Should a wealthy parent move from a mansion into a tiny shack? Should parents replace leather couches with wooden benches? Should parents never go to nice restaurants, and instead eat only fast food?

Fortunately, we can address this problem more elegantly. How can you create the underlying motivation to be financially independent? Not by removing your luxuries. By removing *the child's* luxuries.

Adults often forget what luxuries are for kids. When adults think of luxuries, we might think of oysters, caviar, traveling, Swiss watches, or fancy clothes. None of these mean anything to most kids. Kids' luxuries are video games, junk food, TV, and freedom. Permission to go to a movie with their friends is often more important than what kind of car brings them there.

In fact, many of a child's "luxuries" do not cost anything at all; often children whose parents are not particularly wealthy have luxuries that other kids can only dream about. Being allowed to watch TV until 3 a.m. is an amazing luxury for a child; you do not have to be rich to let your child stay up late.

By removing a child's luxuries, a parent can provide a powerful incentive for the child to work hard. A child lacking specific luxuries has something to strive for, and a way to someday improve his quality of life.

However, it can be emotionally difficult to remove a child's luxuries. This can be particularly true for successful parents who themselves lacked luxuries growing up, and felt that lack acutely. Parents often want their children to enjoy the luxuries that they never had.

Thus, a parent must have a certain discipline to prevent the child from becoming unmotivated because of an excess of assured luxury. It is always easier to not buy an item than to take it away. However, it is much better to take away a child's video game system than to take away his chance of future success.

The luxuries that need to be removed in most cases are video games, TV access, fashion items, cell phones, private phone lines, and computers. While this may seem extreme to some, by restricting access to these you are giving your child the eventual ability to get these luxuries for himself.

Nearly all of the successful people with whom I have discussed this topic have told me stories of specific luxuries that they lacked as children. Sometimes these were luxuries that their parents could not afford. Just as often, their parents deliberately denied them these luxuries. For example, they were not allowed to watch TV, they were not given a video game system, or their parents would not buy them certain shoes. They grew up with strong incentives to develop their own abilities.

My own parents used this technique masterfully, refusing to allow me some luxuries and claiming that they could not afford others. For example, I grew up believing that my parents could not afford cable

TV (my mother is a doctor and my father is an engineer). While the ethics of this are questionable, the effectiveness is unarguable; by age 22 I owned my own highly successful education business. I grew up knowing that to get the things I wanted, I would have to earn them myself. No toy, privilege, or luxury that my parents could have given me could have been even a thousandth as valuable as that incentive.

COMPUTERS

Computers are on the list of luxuries that a parent should remove. This may seem odd; after all, isn't the computer an important tool both in education and in the workplace?

Let's look at the computer myth. The myth says that kids need to use computers daily so that they can learn to use them. Any job will require computer proficiency, so education should incorporate computer use. Children should start using computers early and constantly, so that they can develop the subtle cognitive skills required to use them. Using a computer is like learning a language: a child should start young.

I have worked with students with almost every cognitive deficiency imaginable; I have worked with students with every type of recognized learning disability, and even students with cognitive difficulties so specific that they do not yet have a name. But I have yet to see a single student who is unable to use a computer. Decades ago, computer use may have required some special skills. But modern computers have become so easy to use that using them requires no advanced cognitive skills whatsoever.

Sometimes, adults who grew up without computers are afraid

to use them. They assume that using computers requires some important ability that they lack, and want to make sure that their kids develop that ability. In reality, however, this is just fear of the unknown. A child who is isolated from computers for his whole life will not lack any cognitive skills, but merely a familiarity that can be gained in a day or two.

An even more absurd version of this fallacy is the idea that personal Internet access is an essential requirement for modern education. The idea is that a child with his own internet access will be able to research and learn about science and history.

There is a kernel of truth to this belief. The internet does make certain types of research quicker and easier; however, students usually use the Internet for research once or twice per year. The rest of the time, children usually use the Internet for entertainment. They chat with friends, play online video games, watch online cartoons, visit adult themed websites, etc. A computer with an Internet connection can provide as much mindless entertainment as a TV and video game system combined.

Computer programming, on the other hand, does develop an important system of reasoning. Those who take even basic level courses in computer programming can develop cognitive skills that can be applied to a variety of situations. (Computer programming courses include courses in C++, Java, and other programming languages. These should not be confused with computer applications courses, which cover the use of programs like Microsoft Word.)

The computer myth also states that because computer literacy is required in most jobs, a child needs to have constant computer access. There is some truth to this; almost every job requires a computer. Whether you take orders at McDonald's or guide the strategy of a billion dollar corporation, you will need to use a computer.

And that is the whole point. Whether you work at a fast food restaurant or run a corporate empire, you will use a computer. Clearly, there are things other than computer proficiency that matter. Problem-solving skills, discipline, and creativity make the difference, and those skills should be given priority.

I believe that people should be computer proficient by the time they are eighteen. But they should not be computer dependent. A person should be able to spell without spell-check and add without a calculator.

Thus far, I have only met one person who was 18 and not computer literate. That same person later graduated from Brown University with honors in computer science. The problem-solving skills she had developed during the first 18 years of life mattered much more than the fact that she did not know how to use a mouse until she started college.

This is not to say that you should remove all computers from your house. However, an unmotivated child should definitely not have his own computer. Under no circumstances should he have a computer in his room. He should be allowed to use the computer only to type papers, to perform necessary research, and for other similar purposes. In other words, he should be allowed to use the computer for work, but not for entertainment.

BOYS, TV, AND VIDEO GAMES

If you put a boy in a dungeon with a TV and video game system, he will be blissfully happy. It does not have to be a particularly good

TV or video game system; something in the male mind derives almost limitless pleasure from most video games.

Thus, giving a male child a video game system is the equivalent of giving him about a thousand toys. A single video game system (that includes a computer that can be used to play video games) can demolish a boy's incentives.

Effectively removing a male child's luxuries requires removing all access to video games. Do not worry – he will still find a way to play them (at school, at his friends' houses, even on a graphing calculator). All you can do is limit his access.

SHOULD LUXURIES BE USED AS INCENTIVES?

One popular method is to remove the luxuries, and then use the luxuries as incentives. For example, a parent might take away the child's video game system, and then give it back if he gets an A.

This method rarely works, because it has not fundamentally changed the nature of the incentives. The child still sees his parents, rather than his own abilities, as his source of luxuries. A child who knows that his parents can be persuaded to give him back his video game system will work on impressing his parents. His goal is now to meet whatever standard they have set. If the standard is an A, then he will work to get an A. If the standard is a B, he will work to get a B.

Compare this to the student who knows that his parents will never give him the things he wants. His incentives are completely different. His goal is not to get above some arbitrary cutoff. His goal is to be the best. The child who will get a reward for a B has no incentive to get an A. The child who will get a reward for an

A has no incentive to get an A+. But the child who will get no reward at all from his parents is not looking to meet his parents' standards. Instead, his goal is to do his absolute best to develop his own abilities, since he knows that only his own abilities will give him what he wants. He will want to go to a great college not for parental approval, but for his own benefit. He will be training for himself, not for his parents. This may seem like a small difference, but it is all the difference in the world.

Thus, luxuries should not be used as short-term incentives. Instead, the lack of luxuries should be used as a powerful, permanent, and ever-present long-term incentive. The child without luxuries has an unmatchable incentive to develop his own abilities.

LUXURIES AS ALTERNATIVES

A child who grows up in a household without TV becomes a strong reader. There is no mystery as to why this occurs: everyone needs some kind of entertainment. Children who cannot watch TV often turn to books as their primary source of entertainment.

Thus, the presence of luxuries that provide entertainment, such as TVs and video games, has a doubly negative impact. First, these luxuries can destroy a student's incentive to do well. Second, they provide students with an alternative to reading for entertainment, and thus indirectly weaken reading skills.

WHAT ABOUT EVERYONE ELSE?

This chapter focused on an extreme case: unmotivated students in upper middle class households. But what about other kids? What about motivated students? What about students in other households? Is it okay to allow other kids to have luxuries?

Not surprisingly, the answer is almost always no. For a child to be truly motivated to work, he needs to lack some things. He needs to have something to work for, something to strive for. He needs to fundamentally believe that his own abilities will get him what he wants in life in order to have the maximum incentive to develop them.

7: GIRLS VERSUS BOYS

This is a section about stereotypes, and as with all stereotypes, there will be many exceptions. However, understanding the nature of these differences will help you teach math to both girls and boys. You might find that "girl" techniques are appropriate for your son, or that "boy" techniques work well for your daughter. Or you may find that neither are completely appropriate for your child. However, there is a good chance that the gender specific techniques in this chapter will match exactly.

Take a boy and a girl in the same school, with the same grades, the same level of ability, and the same level of understanding. It would be no surprise to discover that the boy thinks of himself as excellent at math, and the girl thinks of herself as horrendous at math.

There are a variety of reasons for this discrepancy. Girls tend to attribute failures to permanent, internal qualities. For example, a girl who fails a test might think, "I failed this test because I am not good at math."

This is an internal attribution. The girl blames herself (instead of some external factor, like a teacher or textbook) for the math failure.

The girl may also think of the attribute as permanent. She believes that she is not good at math in the same way she might believe her blood type is B: it is a permanent and unchangeable feature.

The boy who fails the same test will be more likely to think, "I failed the test because the test was unfair/stupid/ridiculous. The next one will be better."

The boy attributes the failure to the test itself, rather than to his own lack of ability. This is an external attribution. He also views the problem as temporary: "The next test will be better." Thus, a failing grade does not affect his confidence as severely.

With successes, the reverse often happens. A girl who does well on a test might say that the test was easy, while the boy will say that he did well on the test because of his math skill.

Some may argue that boy has the better outlook, but this is not necessarily true. While confidence is important in math, overconfidence can be more dangerous than underconfidence. Overconfident students often lack the motivation to improve; in their minds, they are already as skilled as they need to be.

The most successful students (and the most successful people), tend to attribute failure to temporary and internal sources. When such a student performs poorly on a math test, he thinks, "I did poorly on the math test because I was unprepared, or because my math skills were weak. For the next test, I will train harder, and make sure I am completely ready for the test." A student with this mindset will strive to improve himself.

CHALLENGES AND ENCOURAGEMENT

When I am joking around with a male student, I might present a problem like this: "This problem is probably the easiest problem I can make. However, you will not be able to do it. Why? Because you have the math skills of lettuce." I say this with a smile or a laugh, so the student knows I am joking. However, I have laid down a challenge. At some level, I have forced an "internal attribution": I have told the student that if he is unable to do the problem, it is because something is wrong with him.

I often introduce particularly hard problems like that. The student knows that the problem is much harder than most, and that no one would expect him to be able to solve it quickly. But after that kind of an introduction he will always do his absolute best to figure out the problem on his own. Most males find it almost impossible to back down from a challenge presented like that.

If I introduced problems like that to female students, the results would usually be disastrous. Rather than being baited by the challenge, many girls would be inclined to give up. I learned this fact the hard way. Early in my career I tried to use the same methods on both male and female students. Soon, 95% of my students were male.

Today I would introduce a hard problem to a female student like this: "This problem is pretty hard. But based on the improvement I have seen you make, I think you are now ready to solve it." I have told her that the problem will be hard. Thus, when she struggles, she will not think that she has some mathematical inability. By discussing improvement, I am reinforcing the belief that math abilities can be developed. Thus, if she is unable to solve the problem, she will understand that as she continues to improve, she will eventually be able solve it. In other words, she will think, "I cannot do this problem yet, but I will soon be able to," instead of, "I cannot do this problem and never will be able to."

Note that there is no need to lower the standards for girls. Girls should not be given easier problems, or more help, or anything of the sort. All that has changed is the presentation of the problem.

Similarly, I have seen many female teachers and parents fail with male students because they offer too much encouragement. Telling a male student, "You are doing really well in math," is often the same as saying, "You no longer need to work hard in math." Even an overachieving male student will not reach his maximum abilities in the presence of excessive encouragement. A little encouragement, used sparingly, can be great for a male student. Too much encouragement generally saps the student's motivation and prevents the student from fully developing his cognitive abilities.

Here are a few examples of how I might present problems to male vs. female students. You probably will not want to imitate the presentation exactly, but these examples can help you develop your own presentation style.

For a boy: "This problem is entitled 'A problem that Joe cannot do because his mathematical skills are those of a four year old zebra.'"

For a girl: "This problem is entitled 'A problem that Jane can do, even though the problem is pretty advanced.'"

For a boy: "Only little kids and giraffes need to use calculators for problems like this."

For a girl: "You've gotten so good at math, you can do this

problem without a calculator."

For a boy: "There is obviously no way you can do this problem, although it is too easy for words."

For a girl: "This problem is hard, but your math skills have improved so much that you will probably be able to do it. Now you are an advanced student."

THE LIMITS

The previous section is a bit oversimplified. In practice, you will challenge male students 90% of the time, and encourage them 10% of the time. With female students it is reversed.

Sometimes you have to trust your instincts. Sometimes a boy needs encouragement. Sometimes a girl needs to be verbally challenged.

But be extremely wary of relying entirely on your instincts, especially when dealing with children of the opposite gender. Try the standard methods first, before switching to the reverse methods.

WHEN TO SWITCH METHODS

At the beginning of this section I mentioned that "female" techniques are sometimes appropriate for male students, and vice versa. How can you tell?

Basically, if a male student is approaching math like the traditional female student, you might try switching to female methods. Here is a checklist:

The male student:

- Believes that his problems are internal and unchangeable. He thinks that he is bad at math, and always will be.
- When challenged or mocked, tends to back down rather than rise up.
- When encouraged, works much harder than he does when challenged.

Similarly, there are times when you might try a "male" approach with a female student. Here is a checklist:

The female student:
- Responds to challenges and taunts by working harder.
- When encouraged, becomes complacent and does not strive at all.
- Always blames the test, the teacher, or the textbook for a poor math grade. Never accepts responsibility for a poor math grade.

Again, chances are that you will not need to reverse methods. In almost every case, I use primarily male techniques for males and primarily female techniques for females.

CULTURAL BIASES AGAINST WOMEN EXCELLING IN MATH

Imagine the following introduction to a test, given in a classroom of 16 year old boys. "This is a test of a specific type of reasoning skills. Women and effeminate males tend to do extremely well on this test. Dominant and masculine males tend to do very poorly on it. Most people who are good at math and sports do horribly on this test. People who are successful later in life also do very poorly on this test."

You would end up with a lot of students sabotaging their own test scores. Many would deliberately get problems wrong to prove their manliness. Few, if any, would try their hardest to figure out difficult problems.

In many cultural settings, women face a similar situation. They are repeatedly told that women are not as good at math as men. At some level, they may understand this to mean that feminine women are bad at math, and that only masculine women are good at math. Thus, seeking to be feminine, many female students subtly sabotage themselves. For example, a female student might give up on a problem after 10 seconds when she could solve the problem if she spent another minute. She might not even attempt problems that look difficult. Why? For the student, it is more important to be feminine than to be good at math.

This problem is subtle and pervasive, and has been the subject of several research papers. There are two things to remember when guiding female students. First, remember that although teenage girls want to be feminine, most do not want to be old-fashioned. Suggesting that the idea that math is not for girls comes from the era before electricity and toothpaste never hurts.

But even more important to the success of female students are parental beliefs. When parents believe that their daughters should be better at math than boys, their daughters are more likely to try their hardest. Sometimes parents fear that a woman who is too good at math might scare away men. To some extent this is true. A girl who is good at math will scare away men – specifically men who are insecure, weak, and stupid. It is hard to see how that is a bad thing.

There is also a school of thought that women are biologically less able to excel at math than men. It is possible that, on average, males are better suited to math. It is equally possible that females are better suited to math. When I was at Brown, the top student in the math department was a woman.

What I do know is that the gender of the student is irrelevant to the student's math performance. Training, far more than innate ability, makes students excel at math. Innate ability might make a 1% difference, but the other 99% is up to the parents, teachers, and students.

8: INNATE ABILITY

Here is what you need to know about innate math ability:

1. People have varying natural math abilities. Some people naturally find math easy to learn. Others naturally find math harder to learn.

2. So what?

If you have a slow metabolism, you can either give up on being fit, or you can accept the fact that you have a slow metabolism and train harder than everyone else. If your child struggles with math, you can make excuses for him, and hold him to a lower standard. Or you can accept the fact that your child will have to work twice as hard as his peers, and that if he does so, he will be at the top of his class.

My most successful students, such as the ones who get perfect SAT and SAT II math scores, are not always the ones with the greatest natural abilities. Instead, they are often the students who train the hardest, study math during their vacations, and refuse to let a lack of innate ability hold them back. Sure, they miss a few months of sitting around doing nothing every summer. But they end up surpassing those with much greater "natural" math abilities.

More importantly, these students learn to not let anything hold them back. Everyone faces hurdles in life. Successful people do whatever it takes to overcome them. The child who refuses to give up develops the willpower necessary to face any major hurdle.

The fittest people are those that train the hardest. A natural athlete who sits on the couch and eats junk food all day will lag behind the average person who trains daily. Similarly, the below average math student who trains daily can surpass the student with innate math talent.

9: THE SAT AND THE SAT II

WHAT DOES THE SAT TEST?

To most people, the SAT Math section is a mystery. It only covers algebra and geometry; however students who have taken calculus often get several problems wrong. Students who get A's in honors math classes sometimes do worse on the math section of the SAT than do students who struggle in regular math classes. In addition, the SAT math section does not cover any advanced topics; a student who knows basic algebra and geometry knows every concept and formula needed to solve any problem on the SAT. These basic concepts include the formula to find the area of a circle, the Pythagorean theorem, and how to factor a trinomial. There are no logarithms, no esoteric geometry theorems, and no trigonometry. In fact, most of the necessary formulas are actually given at the beginning of each SAT math section!

What makes SAT math problems difficult is similar to what makes chess difficult. Most people can learn the rules of chess in a few days. But it can take decades to learn to combine the basic moves to create brilliant strategies and tactics. The SAT is similar in that its challenging math problems require students to cleverly combine basic concepts to create a brilliant solution.

This makes the SAT very different from school tests. If most of a class misses a question on a school test, it is often because correctly answering that question required some specific knowledge that most of the class did not know. It may have been a specific formula, a specific scientific fact, or a specific historical date.

However, a challenging SAT math question never requires any highly esoteric knowledge. Instead, it requires true problem-solving ability. It requires analysis and creativity; it requires a kind of brilliance. In this way, the SAT is similar to an IQ test.

This is why nearly every college uses the SAT as a major part of its admissions process. In fact, many investment banks and management consulting firms, looking to draft the elite right out of college, look at SAT scores. It is a rare test that measures analytical reasoning and problem-solving skills rather than knowledge and diligence.

The SAT score consists of a math score, a critical reading score, and a writing score. Each of these three scores is out of a possible 800 points. Thus, a perfect math score is an 800. Adding the three scores gives the total score, which is out of 2400 points. A student can take the SAT multiple times.

HOW COLLEGES USE THE SAT

Most colleges look only at the student's highest SAT scores, not at his average SAT score. In fact, they look at the highest math score, the highest critical reading score, and the highest writing score from all of the tests that the student has taken. Scores from different tests can be combined. For example, a high math score on one test, a high reading score on another test, and a high writing score on a third test can be combined.

Most competitive colleges use a formula to combine a student's SAT score and grade point average to create an academic selection index. If this number is above that school's cutoff value, the student's application is then considered. If the student's index is below the school's cutoff value, then the application is not considered.

For the most competitive schools (Ivy League colleges, Stanford, etc.), it usually makes very little difference how far above the cutoff a student's SAT score is. In other words, a student with an SAT score of 2390 does not have an automatic advantage over a student with a 2290 (assuming that both scores place the students above the cutoff). The college application essay, teacher recommendations, and extracurricular activities will ultimately determine whether or not the student is accepted.

For most colleges, however, a higher SAT score will give a student an automatic advantage. Most schools will strongly prefer a student with a 2390 to a student with a 2290. Harvard has no trouble finding students with SAT scores above 2300. Most schools do, and will actively seek those students.

THE PSAT

In October of the child's junior year, he will take a PSAT (Preliminary SAT) at school. You do not have to sign up for this; it will be given by the school automatically.

The PSAT is very similar to the SAT. It is shorter, but the subject matter is the same, and it tests the same reasoning skills in the same way. There is no need for specific PSAT training. Instead, SAT training should start before the PSAT.

The PSAT is also used by the National Merit Scholarship program. To become a National Merit Semifinalist, a student needs to get a PSAT score above a certain level, which varies depending on the year and the state of residence.

Below the Semifinalists are the National Merit Commended Scholars. These are students who do well on the PSAT, but do not make the Semifinalist cutoff.

National Merit Semifinalists can become National Merit Finalists and National Merit Scholars based on SAT scores, grades, and recommendations from the school.

Unless a student is a Commended Scholar or Semifinalist, the PSAT scores are not considered by college applications committees. For most students, the PSAT is nothing more than practice for the SAT.

SHORTCOMINGS OF THE SAT AND ALTERNATIVES TO THE SAT

Like any test, the SAT is not flawless. At times, students who possess a high level of cognitive ability do poorly on the test. The reverse can also be true; students with strong reasoning skills may perform poorly on the test.

However, in my experience, the SAT is usually extremely reliable. From time to time, one of my students will get a score that does not seem to reflect his ability. But the vast majority of students with whom I have worked have earned scores that accurately reflect their knowledge and ability.

The primary alternative to the SAT is the ACT, because most colleges accept ACT scores in lieu of SAT scores. The ACT covers similar material to the SAT, but has a very different ideology. Instead of aiming to test reasoning like an IQ test, the ACT tests what a student learns in school. While a difficult SAT math problem requires cleverness, a difficult ACT math question usually requires the student to know how to use an advanced formula. Students who do well in school tend to do well on the ACT.

It never hurts to take the ACT in addition to the SAT. If a student does much better on the ACT than on the SAT, he can use that score for his college admissions process.

THE "TEST-TAKING SKILLS" MYTH

When a child really struggles with standardized reasoning tests like the SAT, it can be very tempting to attribute a low score to a weakness in test-taking skills rather than to a deficiency in cognitive ability. After all, it is much more comforting to believe that a child's low score stems from test-taking anxiety, confusion in the face of multiple choice questions, a tendency to "overthink" problems, or a tendency to make careless mistakes than to believe that the low score indicates major weaknesses in problem-solving and analytical ability. Additionally, the fact that so many mass market SAT preparation programs focus primarily on test-taking skills encourages the belief that what determines SAT success is mastery of test-taking skills rather than knowledge of the subject matter and cognitive ability.

I have trained hundreds of students for the SAT, ranging from struggling to advanced students. So far, only three students have had major difficulties because of weaknesses in test-taking skills (although hundreds of parents initially believed that their children had such weaknesses).

Major weaknesses in test-taking skills are rare. While I have had a few students with weak test-taking skills, thus far I have had only one student with a severe deficit in test-taking ability. He would quickly get the right answer to each multiple choice math problem on the SAT. He would then think that the problem was too easy, and that he must have missed something. He would erase his correct answer and then randomly choose another answer. Despite his excellent math ability, his SAT math scores were low until he learned to stop second guessing.

If your child has a low SAT math score, and you feel that the cause may be a weakness in test-taking skills (rather than in mathematical ability), you can test this very easily. Buy the College Board's *Official SAT Study Guide for the New SAT* or another SAT book that contains practice SAT tests. Each practice SAT test contains a math section that does not have any multiple choice problems. In this section, which contains a total of eight problems, the student has to write in the answer. Have the child do a total of three such sections, and to show all of the steps he took to get the answer. Basically, the child will do 24 SAT questions in which he has to write his own answer. Do not worry about time yet: give the child unlimited time to work on the material.

If the child's low score comes from a difficulty relating to taking multiple choice tests, rather than from a lack of knowledge, he should get nearly every question right. He should get no more than three out of the twenty four questions wrong.

If the child's low score comes from carelessness, he will have written out a solution to every problem. The mathematical steps will make sense, and there will be approximately four mathematical steps for every problem. If his problem is carelessness, he should get some easy questions wrong (the answer key will state the difficulty level of the problem.) If he primarily gets hard problems wrong, then the issue is probably not carelessness, but instead an inability to figure out the difficult math problems.

If the child's low score comes from a significant lack of ability, several problems will have been left blank. Some might have random guesses written; others may have mathematical steps that are irrelevant or illogical. In these kinds of situations, you are not dealing with a problem in taking tests; you are dealing with a deep-seated mathematical deficiency.

TYPES OF SAT PREPARATION

There are two types of SAT math preparation: the cognitive approach and the standard approach. The cognitive approach involves developing the student's knowledge and cognitive abilities. The process is relatively difficult and time consuming, but yields the best results, both in terms of score improvement and in terms of the child's cognitive development. Building these skills through SAT preparation can give students with average math ability the same mathematical reasoning skills as students with excellent math ability. I have had students whose math grades have gone up two letter grades from SAT preparation alone.

The main drawback of these methods is that they are more difficult to use. Full use of the techniques requires excellent math skills, a thorough understanding of how the mind develops mathematical reasoning skills, and a great deal of patience. However, this chapter introduces versions of these techniques that many parents are able to use.

The other type of SAT math preparation is used by most mass market SAT courses and most standard SAT tutors. Instead of building the student's reasoning skills, these courses focus on using the student's current reasoning skills more effectively. These teaching techniques are much faster and much easier to use. In fact, many SAT programs that use these techniques actually hire college students to teach their courses.

If your goal is to spend several months training your child to get a score that will make him a serious candidate at a competitive college, try the cognitive approach. If you want to spend a few weeks preparing your child for the SAT, consider the standard approach.

THE COGNITIVE APPROACH

If you live near New York City, Boston, Los Angeles, or Washington, DC you may have access to private tutors and even group courses that use the cognitive approach. However, the cost of these programs can be prohibitive for many, and there are several ways that a parent can provide similar training.

For best results, start the process before September of the student's sophomore year, at the latest. First, you will need some

SAT math questions to use to teach your child. I recommend getting the Official SAT Study Guide for the New SAT, produced by the makers of the SAT, and *Barron's How to Prepare for the SAT*.

Now you are ready to begin the long and rewarding process of building your child's math reasoning skills through SAT training. The goal of this process is to get a perfect, or almost perfect, SAT math score. Most of the focus will not be on test taking strategies, but instead on developing the reasoning skills necessary to get a top SAT math score.

Begin the process by having your child do math problems from a practice test. The child should not use a calculator at all. In addition, he should not be allowed to see the answer choices in the multiple choice sections. For example, if the question is

2+2=?

A. 1
B. 2
C. 3
D. 4
E. 5

You will cover up the answer choices using a piece of paper. Thus, the student will see

2+2=?

And will not see any of the answer choices. He will have to solve the question directly. Over a period of several months, he will develop a sharper ability to figure out SAT math problems using efficient and direct methods, rather than inefficient methods based on a process of elimination.

Additionally, covering up the answer choices will prevent students from trying to randomly guess the answer. Instead, they will be forced to attempt to figure out each problem. As discussed in Chapter 4, the mind takes the path of least resistance. Covering up the answer choices makes sure that the path of least resistance is the one that best develops the child's reasoning skills.

The student should not be allowed the use of a calculator during any part of the SAT training process. When he does math by hand, he will develop key math reasoning skills as he searches for the most efficient ways to perform math calculations. Keen to avoid extra calculations, he will look for shortcuts and patterns. The ability to find these will be necessary when solving more advanced problems. The following SAT problem illustrates this principle.

$xy=6$

$x+y=9$

Find x^2y+xy^2.

The student who uses a calculator may be tempted to guess different possible values for x and y, and see if they work. He might attempt to solve the problem by solving the second equation for y, and substituting into the first equation, as follows:

$x+y = 9$

Subtract x from both sides of the equation to get

$y = 9-x$

Substitute into $xy = 6$ to get

$x(9-x) = 6$

distribute to get

$9x - x^2 = 6$

subtract 6 from both sides of the equation to get

$-x^2 + 9x - 6 = 0$

Solve for x using the quadratic formula. Then solve for y by substituting the correct value of x into one of the equations. Then substitute the correct values for x and y into the equation x^2y+xy^2 to get the answer.

The student who does not have a calculator will be less inclined to use such a technique, since it involves several unpleasant calculations. Instead, he will look for an efficient way to solve the problem. He may even see the best way to solve the problem:

First, notice that x^2y+xy^2 equals $xy(x+y)$. We are given that xy = 6 and that $x+y$ = 9. $xy(x+y)$ = 6*9 = 54, which is the correct answer.

Removing the calculator option gives the child a strong incentive to solve each math problem in an efficient way that minimizes difficult calculations. He will look for simple, elegant, and brilliant solutions. The ability to find such solutions dramatically increases a student's chances of getting a perfect SAT math score.

Finally, note that some problems require that the student look at the answer choices. This has nothing to do with difficulty; some questions make no sense without the answer choices. For example, look at the following realistic SAT problem:

All of the following have the same value except:

A. $16*1$
B. $8*2$
C. $4*4$
D. $10*6$
E. 2^4

If you cover up the answer choices, this problem makes no sense. Thus when you come to questions like this, you should obviously uncover the answer choices. However, for other questions, including difficult questions, cover up the answer choices.

THE STANDARD APPROACH

The standard approach focuses on test-taking strategies, rather than on building cognitive skills. In math, the standard approach emphasizes process of elimination, "backsolve," and "plug-in."

PROCESS OF ELIMINATION ON THE SAT

The SAT is designed to differentiate a broad range of ability levels. To do this effectively, each question must be reliable. That means that if average students tend to get a problem wrong, below average students should also get that question wrong. Similarly, if average students usually get a problem right, then an advanced student should get the problem right as well.

That means that a person with an extremely low IQ should not be able to solve any problems, since if he were able to solve the problem, then everyone above him should also be able to solve the problem. Thus, every person who took the test would be able to solve the problem. In other words, the question would be pointless; it would not help differentiate smarter students from less capable ones.

Thus, if you can figure out what a person with an extremely low IQ would do, you can often eliminate that answer choice. But how would you know what a person with an extremely low IQ would do?

A person with a low IQ basically thinks like a young child, not like a lunatic. So just ask yourself what a Five year old would do, and eliminate that answer.

The following is the realistic SAT problem discussed earlier:

All of the following have the same value except:

A. 16*1

B. 8*2

C. 4*4

D. 10*6

E. 2^4

If a five year old looked at this problem, he would guess that the answer is E, since it looks different. Thus, without doing any math, you can eliminate choice E. The correct answer is D.

Of course, to solve this problem, you have to use a mathematical process of elimination. You have to determine the numerical value for each answer choice, and see which one has a different numerical value. But when standard SAT courses discuss "process of elimination," they usually do not mean this kind of mathematical process. Instead, they are basically referring to guesswork: the student uses his intuition to eliminate two or three answer choices, and then guesses one of the remaining choices. Obviously, such methods are unreliable.

I personally spend very little time discussing such methods with my students, since my goal is to enable them to solve any problem quickly and efficiently. From my perspective, if a student needs to use a non-mathematical process of elimination, he does not really know how to solve the problem. Instead, I make every student solve every math problem directly.

UNDERSTANDING THE "PLUG-IN" METHOD

Let's look at the following problem:

x^5/x^4

The traditional way to do this problem is to just subtract the exponents, and realize that the answer is x^1, which is the same as x.

However, there is another way to do this problem, known as the plug-in method. We begin by assigning an arbitrary number to x (or "plug in" a value for x). Let's say that x=3. Then we get $3^5/3^4$, which is the same as 243/81, which equals 3. Since 3 is the same as x, we can say that the answer is probably x. However, when using the plug-in method, you can never be 100% sure. The answer could be x. Or it could be 2x-3. It could be x^2-6 for that matter. All you know for sure is that when x is 3, then x^5/x^4 is also 3. To be more certain, you can plug in a few other numbers to see if you get the same result.

The plug-in method can be used to simplify more complicated expressions as well. For example, suppose you have the problem $x^2y*x^3y^4$. We can pick numbers for x and y. Let's say that x=3 and y=2. Then above expression becomes $3^2*2*3^3*2^4$ = 9*2*27*16 = 7776. Through trial and error, one can determine that this is the same as 3^5*2^5. Thus, $x^2y*x^3y^4$ is probably the same as x^5y^5. Again, you cannot be sure, but you can make a reasonable guess.

The plug-in method is popular with some students, because it allows them to do arithmetic instead of algebra. In other words, they can do the problems with numbers rather than with letters.

At a cognitive level, it makes the mathematical thinking more concrete and less abstract. The student is able to replace abstract elements (the variables), with concrete ones (the numbers). Because concrete reasoning is easier, many students prefer to do several extra steps of concrete reasoning in order to avoid abstract reasoning.

CRITICISMS OF THE PLUG-IN METHOD

The most common criticism of the plug-in method is the risk of error. For example, look at the problem 3x+2x. The correct answer is

5x. However, suppose the student plugs in 0 for x. He will then get $3(0) + 2(0)$, which is the same as 0. Since x=0, the student may think that the answer is x, instead of 5x.

This type of error is common; however, it is caused by using the plug-in method incorrectly. To be safe, the student should test at least two completely arbitrary numbers. For example, he can first try using 7 for x, and then try using 3 for x. The student will probably be able to figure out the correct answer. Thus, if the plug-in method is used correctly on an easy SAT problem, the risk of making this specific type of error is relatively low.

However, because the plug-in method requires many extra calculations, the probability that the student will make at least one mistake increases. A student has a higher chance of making a mistake when he is doing twenty mathematical steps than if he does only three mathematical steps.

Another criticism of the plug-in method is the amount of time it takes. Because using the plug-in method requires many, many extra steps, it takes much longer than an algebraic method. Thus, a student with a reasonably strong grasp of algebra is often wasting time when he uses the plug-in method.

This objection is valid. It makes little sense for a student who is skilled at algebra to use the plug-in method. (However, a student with weak math skills is in a different situation. Using the plug-in method, a weak math student may be able to solve a problem that he would never be able to do algebraically. It might take him a long time, but at least he will get the answer.)

The final criticism of the plug-in method looks at its cognitive impact. Each time a student uses the plug-in method, he is using concrete reasoning instead of abstract reasoning. Every time he does this, his algebraic skills lose the opportunity to develop. Over time, his abstract reasoning and algebraic skills may atrophy.

This becomes an issue when the student faces a more difficult SAT problem. For example, the student may be given a problem like this (source: *The College Board's Official SAT Study Guide for the New SAT*):

$xy = 7$. $x-y = 5$. Find $x^2y - xy^2$.

This problem cannot be done using the plug-in method. There are several ways to do it, but the most efficient way is as follows. First, factor the expression $x^2y - xy^2$, to get

$x^2y - xy^2 = xy(x-y)$. Since $xy = 7$ and $x-y = 5$, that is the same as $7*5 = 35$.

The issue with the plug-in method is not that it cannot be used to solve the problem; no method will work in every situation. The real problem with the plug-in method is that it prevents the student from developing the cognitive skills necessary for solving that problem algebraically. By giving the student the incentive to let his abstract reasoning and algebraic skills atrophy, the plug-in method has made it much more difficult for the student to learn to solve more advanced problems.

If a weak math student is going to do only three weeks of preparation and then take the SAT, it may be a good idea to introduce the plug-in method. It is an okay last-minute strategy to know about.

But more and more students are spending several months training for the SAT; in fact, many are spending more than a year. When a student uses the plug-in method for an extended time period, the results can be disastrous. Their algebraic skills can atrophy to the point that they cannot begin to approach any difficult algebraic problem. Months of retraining may be necessary to rebuild the damaged skills.

Thus, I never teach the plug-in method or let any of my students use it. Compared to a direct, algebraic approach, the plug-in method is inefficient and ineffective.

THE BACKSOLVE METHOD

The "backsolve" method only works on multiple choice problems. Most SAT problems are multiple choice, but each SAT also has several problems that are not multiple choice.

Look at the following problem:

$3x-8=10$. Solve for x

A. 2
B. 4
C. 6
D. 9
E. 12

The traditional way to do this problem is as follows:

$3x-8=10$

add 8 to both sides of the equation to get

$3x = 18$

divide both sides of the equation by 3 to get

$x=6$

This problem can also be done by testing out each answer choice. Basically, you can try out each number for x, and see if 3x-8 ends up equaling 10. We can start with choice A, which is x=2. The left side of the equation becomes

$3(2)-8$, or $6-8$, or -2.

This does not work, so let's try choice B:

$3(4)-8 = 12 - 8 = 4$.

Again, this does not work, so we try C next.

$3(6)-8 = 18 - 8 = 10$

Since this gets the desired answer, we know that the answer is C.

Although hard SAT math questions can rarely be solved using the backsolve method, it can be handy for some questions of medium difficulty. However, because it is much less efficient that direct methods, I do not teach this method or let my students use it.

ADDITIONAL RESOURCES

Often students, tutors, and parents are looking for detailed explanations to the practice tests in the Official SAT Study Guide for the New SAT. There are two excellent sources.

1. SAT Math Cognition (www.SATMathCognition.com). This is an e-book available through my company, AVE. It has detailed walkthroughs for all the questions in the practice tests of the Official SAT Study Guide for the New SAT. Free samples are available online.

2. The Official College Board Online Course (www.Collegeboard.com). This online course contains answer explanations as well, although they are written in the style you might find in a college math textbook. If you are excellent at math, they can be helpful. Additionally, the Online Course has extra practice tests.

THE SAT II

In addition to taking the SAT I, students applying to competitive colleges will need take two or three SAT II subject tests, depending on the specific college's requirements. These are hour-long tests on specific subjects, such as US History, Biology, Literature, or Physics. The maximum score on a subject test is 800, and the minimum is 200.

There are two math SAT II subject tests, the Level 1 and the Level 2. The Level 1 focuses on algebra and geometry, while the

Level 2 focuses on precalculus. Although the Level 2 has more difficult problems, it also has a more generous curve. A person can get a few questions wrong and still get an 800 on the Math Level 2 test.

I recommend that students applying to Ivy League colleges from competitive areas (New York, Boston, Washington, DC, etc.) take the SAT II Math Level 2 test (as long as they have studied precalculus). Although either test can be used for college admissions, the Level 2 test is often considered more favorably than the Level 1 test.

Thorough preparation for the SAT II is more straightforward than thorough preparation for the SAT I, since it is primarily a test of knowledge rather than a test of intelligence. While the SAT is like an IQ test, the SAT II is more like a school test.

While a child's school will provide some preparation for the SAT II, few schools provide sufficient preparation. Fortunately, there are several options for outside preparation.

My personal favorite preparation books are Barron's *How To Prepare for the SAT II Math*. These prep books contain practice tests with problems that are generally a bit more difficult than the actual test. These problems can thoroughly prepare students to succeed on the SAT II. Barron's offers two books, one for the Math Level 1 test, and one for the Math Level 2 test.

Other companies also make decent SAT II prep books. Princeton Review's practice problems tend to be a bit easier than the problems on the actual test, so I rarely use them. Kaplan's SAT II prep books are also okay, and tend to be a bit more challenging than those from the Princeton Review. You can find particularly difficult questions in Rusen Meylani's practice test books, and I sometimes use these books with advanced students to help ensure perfect scores.

TUTORS

A good tutor can make the SAT II training process more efficient and effective. If time is a major issue (for example, if your child starts preparation during September of his senior year), you may want to strongly consider getting a tutor. If your child starts before March of his junior year, he may prefer to try independent preparation with a test prep book first.

10: QUIZ VERSUS EXAM PERFORMANCE

Often, a child will do extremely well on math quizzes, but horrendously on exams. The student's quiz grades may be the highest in the class, and yet he may be getting D's on his exams. This is not due to a lack of studying – the student studies for days for each math exam, putting in far more effort than his peers. In fact, many of his peers do not study at all for math exams, and still do much better than he does. It seems that no matter how much he studies for an exam, he does horrendously.

HIERARCHIZATION REVISITED

The cause of the student's difficulties is lack of effective hierarchization. Rather than remembering the most important concepts for a long period of time, he is memorizing every single detail for a few days. Thus, his quiz scores are excellent; after all, he has memorized every single detail from the given chapter.

Unfortunately, he has memorized dozens or even hundreds of details for each chapter, while his peers have learned only two or three concepts. By exam time, his peers have based their understanding on about a dozen key concepts. The struggling student, however, has based his understanding on thousands of details. In fact, he may not really understand the concepts at all, and may be relying entirely on rote memorization. There is far too much information for him to keep in his memory; no matter how long he studies, he will never be able to keep all of it memorized for the exam.

FIXING THE PROBLEM

Correcting this problem requires a lot of effort from both the student and the parent. The process will reshape the way the student approaches math, and will improve not only his exam performance, but his standardized test performance as well.

We must first create an effective incentive. Because you are teaching a diligent student, you might not need to give the child an incentive to put in effort. However, you do need to create a cognitive incentive. You need to give his brain a strong reason to hierarchize the information, rather than to just memorize every single detail.

First, test him daily on key concepts from his math class. If you happen to be excellent at math, you can just write two or three review problems every day. Otherwise, just pick review questions from previous chapters. If the textbook has chapter review questions or practice tests, choose two or three problems from those sections every day, and have the child do them.

When your child does them, he should not be allowed to look back at the chapter. He should do the problems as if he were taking a test. Why? We already know the child is adept at memorizing information for a short period of time. If he is allowed to look back at the chapter, he will obviously be able to figure out how to do the problems. Our goal is to teach the child to develop a permanent understanding instead of relying on short-term memorization. That means that the child will often struggle, but, as you have already learned, struggle fundamentally develops the child's cognitive abilities. If the child is struggling with a math problem, you are doing your job right.

Additionally, the struggle creates a powerful cognitive incentive for the mind to hierarchize information. If the child remembers the key concepts, he will be done in a few minutes. If he forgets those concepts, he may be struggling for hours.

When selecting review problems, select basic ones. Do not pick the hardest problem in the chapter, since that problem may require the child to remember some specific formula or detail. Your goal is to help the child develop a permanent understanding of basic problem types, not a temporary understanding of some highly unusual problem.

Finally, the pattern of review problems should be unpredictable. Remember, your child likes to memorize details for a short period of time, and you have to eliminate that option. If you give him problems from Chapter 3 on Monday, Chapter 4 on Tuesday, by Wednesday he will have memorized Chapter 5. The problem is, of course, that he will have forgotten Chapter 5 by Thursday. He is not doing this maliciously – that is just what his brain is used to doing.

To be more effective, on Monday, give him one problem from Chapter 3, one from Chapter 8, and one from Chapter 1. On Tuesday, give him one problem from Chapter 4, one from Chapter 3, and one from Chapter 17 (notice that Chapter 3 reappeared. The child who prefers memorization may assume that Chapter 3 will not reappear on Tuesday, and may therefore forget the Chapter 3 concepts.)

Give the child no option other than to hierarchize the information effectively. Make him develop a permanent understanding. In a few months you will have a happier child who is better at math, and better at taking math exams.

COMBINING INFORMATION

Most questions on math quizzes focus on only one concept. However, exam questions often require students to combine several concepts to solve a single problem.

Often, students who do well on quizzes and poorly on exams completely compartmentalize information. They view each concept as distinct, and do not look for connections between different concepts. Thus, when faced with an exam question, they attempt to find a single concept or formula that will allow them to solve the problem. Because the exam questions combine several concepts, they baffle such students.

Thus, you need to give your child practice problems that require using multiple concepts. If you are great at math, you can write them yourself. Otherwise, I would recommend using problems from SAT or SAT II practice tests. Have the student do 3-5 problems per week.

As the student learns to combine the information and to hierarchize it, he will be making the important transition from a diligent math student to a truly excellent math student.

THE OPPOSITE PROBLEM

Some students have the opposite problem: they do horribly on quizzes and well on exams. Fortunately, this problem is usually easy to solve. In almost all such situations, the student has not been doing his homework every day. Make sure the student does his homework daily and quiz him on the main formulas; his quiz grades should dramatically improve. It is even better to take this one step further, and to make sure the student does his homework a week before it is due.

11: THE CALCULATOR FALLACY

The most significant change to American math education during the last fifteen years has been the incorporation of calculators. As early as seventh grade, students are using graphing calculators as part of their math education.

You already know that for most students, math is valuable not for its potential applications, but for the cognitive abilities and reasoning skills it develops. For example, a student should study geometry so that he can develop his logical and spatial reasoning skills, not because he will one day need to decide whether or not two triangles are congruent. A student should learn to factor polynomials because of the cognitive skills that the task develops, not because he is likely to use this skill in any part of his adult life.

However, many teachers and administrators do not fully appreciate this fact. Instead, they believe that the goal of math education is to give students the ability to somehow get the answer to a large number of math problem types. Because calculators can often allow students to get the answer more quickly, many teachers teach students to solve problems using calculators. For example, rather than teaching analytical techniques for graphing, teachers just teach students how to graph using a calculator. In the absolute worst cases, teachers actually teach basic math concepts, such as adding and multiplying fractions, with calculators.

Of course, this approach completely misses the point of math. The answers to the problems themselves do not matter; only the cognitive skills the problems develop matter.

Using a calculator to graph $y=x^2+1$ is of little value. A student who does so is not developing his reasoning skills, but is just learning to enter formulas into a calculator.

Proponents of calculator use often point out that calculators allow students to study more advanced topics in math. This argument is based on a lack of knowledge of what advanced math actually is. In college, I studied abstract algebra, calculus of manifolds, real analysis, number theory, differential equations and other advanced topics. None of those subjects involved any calculator use whatsoever. Neither do advanced courses in topology, complex analysis, or any of the other major areas of advanced abstract mathematics. In fact, many advanced math topics do not even use numbers.

However, there are certain types of calculations that are extremely unpleasant to do by hand. These problems are not necessarily the most advanced, but they do require either a calculator or a lot of free time. For example, regression analysis, which can take hours by hand, can be done in a few seconds by even a fourth grader with a graphing calculator.

This can make the course of study sound more impressive. "Cubic regression analysis" sounds a lot more advanced than "adding fractions," and often impresses parents. The problem is that the students are not actually studying regression analysis; they have just memorized four calculator steps. Instead of developing their reasoning abilities, they have memorized how to use a calculator to solve a problem that they will probably never face in adult life.

Other supporters of calculator use argue that using a calculator allows a student to focus on the more important parts of the problem without having to worry about things like addition and multiplication. This argument sounds credible, but it misses much of the reality of the situation.

In order to focus on the more "important" parts of the problem, you must be able to understand the required patterns and relationships. This requires the presence of certain math abilities, and chronic calculator use causes these fundamental math skills to atrophy. For example, I have worked with hundreds of high school students who could not add fractions by hand. This becomes a problem when the student studies more advanced topics. For example, there is an obvious correlation between the skills used in adding $1/2$ and $1/3$, and the skills used to add $1/(x+1)$ and $1/(x+2)$. A student who has forgotten how to do the first problem by hand will find it more difficult to learn how to do the second problem.

But the damage can be much more subtle. When you do math in your head, you may look for ways to do the math more easily. For example, if you multiply 66*15 in your head, you might realize that 66*15 is the same as 66*10 + 66*5. 66*10 is 660, and you might realize that 66*5 is just half of that, or 330. Thus, 66*15 is 660+330 = 990. Thus, multiplying 66 by 15 becomes a creative process and a mental workout. It reinforces your mathematical abilities and keeps your mind sharp.

On the other hand, multiplying 66 by 15 on a calculator is a mindless task. It develops no cognitive skills whatsoever.

Now apply that principle to the thousands of calculations that a student does in school. Each calculation done mentally or by hand is another chance to approach math cleverly and creatively, while keeping the student's cognitive skills sharp. It begins to add up quickly. It is no coincidence that in other countries, where students are not allowed to use calculators, the students outperform American students in math (although those countries spend far less per student on education).

Proponents of calculator use may argue that not using a calculator is like walking instead of driving: sure, it might keep you fit, but it slows you down so much that it is not at all practical. This argument misses the most important point. When you are driving somewhere, your goal is to get there. If you want to go to the bookstore, your goal is usually to get there as fast as possible; a car is the best way to do that.

But when you are studying math, you are exercising your mind to develop your cognitive skills. Your goal is to strengthen your mind, not just to get the answer. The answer is never important. No one cares what the cosine of 30 degrees is, or how long it takes Jane to reach the store. The only thing that matters is the cognitive development that results from figuring out how to solve the problem.

Math does for the mind what lifting weights or running does for the body. Using a calculator is not like using a car to reach your destination faster. It is more like exercising by driving a car around a track instead of running.

In fact, the belief that calculators help students work faster is often completely false. Instead, students who do math without calculators often end up solving problems more quickly than other students. A student solving a complicated problem spends very little time doing actual calculations. Most of the time is spent examining relationships and determining what concepts apply. The student who does math by hand has these concepts ingrained in his mind, and is adept at using them. Thus, he rapidly sees relationships between various formulas and concepts, and can quickly figure out how to do the problem. On the other hand, the student who relies on the calculator sees the relationships slowly, if at all, because the concepts are not thoroughly ingrained in his mind. Thus, while the strong math student figures out how to solve the question in a minute, the calculator-dependent student may spend an hour and still not figure out what to do.

Additionally, the strong student can also do many calculations faster than the calculator-dependent student. For example, by the time the calculator-dependent student types 77*10 on his calculator, the strong math student already knows the answer is 770. He has every incentive to know the fast and efficient ways to do a problem, and is therefore always ready to use them.

HOW TO BE THE BEARER OF BAD NEWS

It is never easy to tell people what they do not want to hear. It is even harder when you are opposing an exceptionally convenient belief.

Sure, teaching math with a calculator does not work that well. But what if it did? What if you really could learn math just as well with a calculator, and would never have to multiply decimals or add fractions by hand? Wouldn't that be great?

While we are dreaming, wouldn't it be great if we all had magic lamps that made all our wishes come true?

I think that at some level, most parents and math teachers know that teaching math with a calculator does not work. But it is so much easier than teaching the right way. When faced with a student who just does not get a concept, it is very tempting to just teach a few calculator keystrokes instead of spending hours teaching concepts and developing cognitive skills.

In many schools, there are some teachers who do not let their students use calculators. They are usually the best teachers in the school, both in terms of knowledge of the subject and in terms of dedication to the art of teaching. But there is a good chance that your child's teacher has incorporated the calculator into the math class. That means that you will be telling your child not to use it against the teacher's instructions, which might not be easy. The child will view the teacher as the authority on math, and you, of course, are "just his parents" (even if you happen to have a PhD in math or physics). And, to make matters worse, you are telling him exactly what he does not want to hear.

It is not easy to convince a student to not use a calculator, but it can be done. Almost all of my students work without their calculators, even when I am not with them. They understand the damage that calculator use causes, and thus refuse to use calculators. In fact, some of my students have even argued with teachers about calculator use, and refused to use them in class. Several of my students have even taken the SAT without calculators; most returned with perfect SAT math scores.

Here are some techniques:

1. Explain the issue in simple terms. Explain that using a calculator is like using an electronic wheelchair when you are

able to walk perfectly – it just makes your muscles atrophy. Let the student know that using a calculator is weakening his mind.

2. Challenge or tease the student in a lighthearted way. This tends to work better with male students. For example, you can suggest to the student that using a calculator is like using training wheels on a bike.

3. When all else fails, just take his calculator away, and watch him do his homework to make sure he does not use a calculator.

People often ask what I do to make sure that my students do not use calculators. Much of what I do will not work for everyone, but some parts may be helpful.

First, I explain the dangers of using calculators, and then I emphasize the idea with extreme and colorful analogies. For example, I tell students that a calculator is like plutonium – it has its uses, but it is dangerous to keep around. If necessary, I then explain what plutonium is (plutonium is used in nuclear bombs. It is highly radioactive, and even a small sample can cause death by cancer or radiation sickness.)

When students mention that their teachers say it is okay to use calculators, I may point to the other times in history when accepted views turned out to be wrong (for example, I may remind the students that there was a time when virtually all political leaders accepted the practice of slavery).

More importantly, I work hard to earn and keep a student's respect. First, I know the subject well. Secondly, I never lose my patience with a student. To understand why this is important, imagine if you went to a brain surgeon who tended to yell and lose his temper. Would you feel safe in his hands? Would you think that you were dealing with a competent and effective brain surgeon?

Next, I stay consistent. There are times when it is tempting to let a child use a calculator. For example, at the end of a long training session, you might be tempted to let your child use a calculator to finish the last problem. Do not give in to that temptation. Times like that are when you show the student how strongly you believe in your principles.

Finally, it is important to practice what you preach. If you tell a child not to use a calculator and then constantly use one yourself, the child will be less inclined to take you seriously. When you work to improve your child's math skills, sometimes your own skills improve as well.

12: THE LADDER, THE RULE OF TWO, THE TIMED LADDER, AND THE HYDRA

A child cannot control the pace of a school math class. If a student does not learn how to do a problem type, he just falls behind. He will later have to struggle to catch up, or just end up getting low grades in the course. The course moves at a fixed pace, and one student's performance will not speed up or slow down the class.

However, when a child is working with a parent, his performance does control the pace of the training. If he moves slowly and is unable to figure out how to solve problems, his parent may move slower and give the child easier problems. The child will end up actually doing less total work.

On the other hand, if the child works quickly and uses his full ability, the parent may increase the number of problems and the level of difficulty. By working at his peak, the student has actually increased his total workload.

Thus, a student has a strong incentive to drag his feet and use only a fraction of his full abilities when learning math from his parents. At some level, he realizes that the worse he does, the less total work he will end up doing. Thus, he might take ten minutes to solve a problem that he could do in one minute, or even give up on problems that he would normally be able to solve. Rather than developing his cognitive abilities, the child ends up wasting his time.

Thus, to ensure that the child gains the benefits of parental math training, the parent must give the child an incentive to push his mind to the limit. Historically, parents have used physical and verbal punishments to create incentives. But what if there was a more elegant and effective system? What if any "punishment" actually benefited the child?

THE LADDER (AGES 8-18)

A parent gives a child a problem. The child gets the problem wrong. The parent should then just give an easier problem, right? For example, if the child cannot add $^{11}/_{23}$ and $^{17}/_{47}$, the next problem should be something like $^1/_3 + ^1/_2$. However, if the child gets the problem right, he is ready for a more difficult problem, right?

Not exactly. The above situation creates the wrong incentive. If the student gets the problem right, he gets a harder problem. If he gets it wrong, he gets an easier problem. Unless he likes doing hard math problems, he has a strong incentive to get the problem wrong. (If your child does like doing hard math problems, consider yourself lucky and move on to the next section.)

The Ladder is a simple drill that addresses the problem. Here are the rules:

1. If the child gets a problem wrong, the parent explains how to do the problem, and then gives the student a more difficult

problem of the same type. For example, if he gets a multiplication problem wrong, he gets a harder multiplication problem.

2. If the child gets the problem right, the ladder ends. Either move on to a new topic, or finish math training for the day.

You can start the Ladder at any level of difficulty. You do not have to start with an easy problem.

The Ladder can be as steep as you want. That means that if the child gets a problem wrong, the next problem can be slightly more difficult, or ten times more difficult. Try different methods, and see what works best for your child.

THE RULE OF TWO (AGES 3-90)

Sometimes a student will have trouble remembering a specific formula. For example, he may have trouble remembering that the area of a triangle is $1/2$ *base*height. Even a student who really wants to memorize the formula may find it difficult to do so. Particularly diligent students may even make flash cards, or write the formula over several times, and may still find themselves unable to memorize the formula.

For a parent, this can be particularly frustrating. After a child forgets how to find the area of a triangle several times, the parent may lose his patience.

Like the Ladder, the Rule of Two gives students the conscious and unconscious incentives to memorize the formula. The Rule of Two is simple. If a student forgets an important formula, he has to write it down twice. If he forgets it again, he has to write it down four times. If he forgets again, he has to write it down eight times. Each time he forgets it, he has to write it down twice as many times as before. This creates a simple and powerful incentive for a student to memorize the formula.

You should use the Rule of Two only on important facts and formulas. These include:

- The quadratic formula
- Multiplication facts up to 12*12. The fact that 6*7 = 42 is an example of such a fact. If the student forgets that, he would have to write down 6*7=42 twice. He would not have to write down the entire multiplication table for 6.
- The area formulas for a circle, rectangle, triangle, parallelogram, and trapezoid.
- The fact that $(a+b)^2 = a^2+2ab+b^2$
- The Pythagorean Theorem.
- Other formulas of similar importance.

As discussed earlier in this book, successful math students hierarchize information effectively. That means that they remember the most important information much better than less important information. Using the Rule of Two only on important formulas helps maintain that hierarchization. Using the Rule of Two on less fundamental formulas weakens this hierarchization.

When using the Rule of Two, you should keep a running total. In other words, you do not start fresh every day. If a student forgets a formula, he writes it down twice. If he forgets the formula again two months later, he writes it down four times. If he forgets it again a year later, he writes it down eight times. A notebook can help you keep track.

THE TIMED LADDER (AGES 8-18)

The Ladder has one weakness that particularly clever and lazy students have found. The Ladder gives a student an incentive to do the problem right; it does not give the student an incentive to do the problem fast. A student who takes a longer time to do a problem may end up having to do less total work than a student who works fast.

To encourage a student to work fast, you can use the Timed Ladder. There are two versions.

Here are the rules for version 1:

1. If the student does not finish the problem in the set amount of time, the parent explains how to do the problem and then gives the child another problem of the same level of difficulty. However, he has less time to do it. For example, he might be given one minute to solve a division problem. If he does not solve the problem in one minute, the next time he only has 50 seconds.

2. As soon as he solves the problem, the Timed Ladder ends.

Here are the rules for version 2:

1. If the student does not finish the problem in a set amount of time, the parent explains how to do the problem and then gives the student a harder problem to do in the same amount of time. For example, if he does not solve a division problem in one minute, he then has to solve a harder division problem in one

minute.

    2. As soon as he solves the problem, the Timed Ladder ends.

THE HYDRA (AGES 8-80)

The Hydra is the drill my students dread the most. In Greek mythology, the Hydra was a many-headed monster that had one particularly annoying habit: anytime you cut off one of its heads, two new heads would grow in to replace it.

The Hydra that I use was inspired by that story. If a student gets a problem wrong, he gets two additional problems of the same type. For example, if he gets a division problem wrong, he gets two new division problems. If he gets one of those wrong, he gets two more, and the process continues like that. Again, the student has a strong incentive to do his absolute best to solve each problem correctly. As soon as the student has solved all of the problems, the Hydra ends.

The Hydra is great when you want to work on many different types of problems, such as during exam review or preparation for a standardized test. I often use the Hydra when preparing younger students for standardized tests like the SSAT and the ISEE (these are standardized tests used for admission into private schools). For example, you can use a math section out of a commercially available practice test book. The whole section (usually 20-25 problems) is the Hydra. If the student gets a problem right, he moves to the next problem. If he gets a problem wrong, he must do two similar problems before he can move forward. When he solves the last problem, he is done. Thus, if the student does well, he has to spend less time working on math. If he does poorly, he may end up spending several extra hours. This gives him a strong incentive to do his best.

Similarly, the Hydra can be used when reviewing for a school test or exam. You can use chapter reviews from the textbook (chapter reviews are usually found at the end of each chapter). Every time the student gets a problem right, he moves to the next. If he gets a problem wrong, give him two similar problems. Once the student solves the last problem in the chapter review, he is done.

A particularly effective drill is the Timed Hydra. This time, the student has a set time (usually one minute) to do each problem. If he does not solve the problem right within the set time, he gets two more problems. He now has the same set time (e.g., one minute) to spend on each new problem. As soon as he correctly solves all problems within the set time he moves on to the next problem.

In fact, the Hydra can be used during any training session. Suppose the student forgets how to add fractions. Give him two more fraction addition problems, and let him know that a Hydra has begun. When I work with a student, anytime he gets any part of any problem wrong, he immediately gets a Hydra focusing on the relevant topic. The Hydra always starts with two problems, but it can quickly turn into dozens of similar problems.

Students develop their minds most effectively when they push their abilities to the limit. The Ladder and the Hydra give students the incentive to do so with every problem they face.

13: THE MICRO-CHALLENGE METHOD

The Asian method can be used by all parents and teachers, regardless of their math ability. The Ladder, the Rule of Two, and the Hydra can be used by most parents. In this section we will examine the Microchallenge method, which can be used by parents and educators with excellent math skills. Since developing the Microchallenge method, I have used it daily with excellent results, even when teaching students with major deficits in their math skills. While the Microchallenge method can be difficult to master, those who can master it will find it extremely effective, especially when combined with the Hydra. It builds a thorough understanding of mathematical concepts, while simultaneously developing many necessary cognitive skills. This method, combined with the Hydra, will allow you to turn students with almost no math ability into strong math students, and to supercharge the abilities of strong math students.

The Microchallenge method requires both patience and a strong understanding of the subject. You should only use the Microchallenge method with material with which you are comfortable. For example, if you are not comfortable with trigonometry, you probably will not be able to use the Microchallenge method to teach trigonometry; instead you might use the Asian System to teach trigonometry. However, you will still be able to use the Microchallenge method to teach the other parts of math.

The Microchallenge method involves helping the student understand how to solve a difficult problem by leading him through a series of "microchallenges" that run parallel to the required cognitive steps of the problem. This ensures that the student develops the underlying understanding and cognitive abilities necessary to do the problem, and does not instead just memorize a few steps.

The following dialogue illustrates how I would use the Microchallenge method to teach a student how to do a specific math problem, and shows both the patience and level of understanding required from the teacher. The example student is indicative of some of the struggling students with whom I have worked.

In this dialogue, the student has to do the following problem:

Add $\frac{1}{x} + \frac{1}{(x+7)}$

During this entire dialogue, the student writes down the questions and answers. For example, if I ask the student to add $\frac{2}{7} + \frac{3}{7}$, the student would write down $\frac{2}{7} + \frac{3}{7}$.

Student: I have no idea how to do this problem.
Arvin Vohra: Yes you do. *<As always, I first try to see if the student can figure out the problem.>*
Student: No, seriously I don't.
AV: Give it a shot – you might be able to.

Student: How do you add something like this? I have never seen something like this before.

AV: How do you add $1/2$ and $1/3$? *<Once I am sure that the student cannot do the original problem, I give him an easier problem that follows the same pattern.>*

Student: Umm… Is it $1/5$? *<At this point, I realize the student probably has no idea how to add fractions with different denominators.>*

AV: Nope.

Student: Is it $1/6$?

AV: What do you need to have in order to add fractions? *<The correct answer is "a common denominator" or "a least common denominator." I ask the student this question for two reasons. First, the question may jog his memory. On the other hand, if he cannot answer the question, at least he will realize that he is missing something important. When he gets the answer later on, his mind will already be primed to incorporate the information.>*

Student: Wait…actually I don't remember.

AV: Yes you do.

Student: Something… I'm not sure.

AV: Can you do $1/7 + 3/7$?

Student: No.

AV: Try.

Student: Is it $4/14$?

AV: Nope.

Student: Is it $3/14$? *<I now see that the student cannot add fractions at all, which indicates that he does not have a visual understanding of fractions. He is not picturing $1/7$ as, for example, a circle cut into 7 pieces with on piece shaded in. Either he is unable to picture $1/7$, or he*

does not create the relevant mental picture for the problem. I now need to see if he has any understanding at all of what a fraction is. In other words, does he know what $3/7$ means?>

AV: Draw a circle *<The student does so.>* Now cut it into seven slices, as if you were slicing a pie or birthday cake. *<The student does so.>* Good. Shade in $3/7$ of the pie. *<The student does so.>* Excellent. Now shade in another $1/7$ of the pie. *<The student does so.>* Good. How much of the pie is shaded in?

Student: Ohhh....$4/7$. So does $3/7$ + $1/7$ equal $4/7$? *<The student is now developing a concrete understanding of adding fractions. Note the importance of giving the student a concrete understanding (using the picture) rather than saying something like "just add the top numbers, and keep the bottom numbers the same." If you do something like that, the student may memorize this as a rote formula, distance himself from the information, and gain no real understanding of the concepts.>*

AV: Yes. How about $1/3$ + $1/3$?

Student: $2/3$?

AV: Yes. Draw a picture of that. *<I reinforce the student's visual understanding of fractions; I do not simply assume that he learned it perfectly the first time.>*

Student: How?

AV: What did you do the last time?

Student: Oh, draw a circle like this?

AV: Good. And then?

Student: Cut it into three parts?

AV: Show me. *<The student does so.>*

AV: Good.

Student: Then shade in one part, and then shade in another part?

AV: Yes.

Student: So it is $^2/_3$?

AV: Good. What is $^1/_9$ + $^4/_9$

Student: $^5/_9$.

AV: Draw me a picture to explain why. < *The student does so quickly and with confidence. He is ready to move forward.* > Good. What is $^2/_{11}$ + $^3/_{11}$?

Student: $^5/_{11}$.

AV: How about…$^2/_{347}$ + $^{11}/_{347}$?

Student: $^{13}/_{347}$?

AV: How about $^7/_{5001}$ + $^{43}/_{5001}$?

Student: $^{50}/_{5001}$. < *At this point, the student is able to add fractions with similar denominators. He is ready to move forward.* >

AV: Good. How about $^3/_4$ + $^1/_8$?

Student: I don't know.

AV: What makes it hard?

Student: The numbers on the bottom are different.

AV: Correct. The number on the bottom is called the denominator.

Student: The denominator?

AV: Correct.

Student: I think my teacher used to talk about denominators.

AV: That makes sense. When the denominators are different, that makes things hard. What would you prefer?

Student: If the denominators were the same.

AV: So make them the same.

Student: How?

AV: $^3/_4$ is how many eighths?

Student: I don't know. It's not $3/8$, is it?

AV: I don't know. Draw me two circles. <*The student does so.*> In the first circle shade in $3/4$.

Student: Should I cut the circle into four parts first?

AV: Yes. <*The student draws a circle, cuts it into four parts, and shades in three.*> Good. Now shade in $3/8$ of the other circle. <*The student does so.*> Is $3/4$ the same as $3/8$?

Student: No.

AV: Which is bigger?

Student: $3/4$ is bigger, obviously.

AV: Do this. See the circle with 3/4 shaded?

Student: Yes.

AV: Cut each of those four slices in half. <*The student does so.*> How many slices do you have?

Student: Eight.

AV: And how many slices are shaded?

Student: Six.

AV: So $3/4$ is the same as what.

Student: Ohh… $3/4$ is the same as $6/8$!

AV: Good. So $3/4 + 1/8$ is the same as…

Student: $6/8 + 1/8$, so $7/8$?

AV: Good. So what do you have to do to add fractions?

Student: Make the denominators the same?

AV: Good. You know what you call it when the denominators are the same?

Student: No, what?

AV: You call it a common denominator. You have to find a common denominator.

Student: Oh yeah! My teacher sometimes talks about that, but I had no idea what she was talking about.

AV: Now you do. So we said that $^3/_4$ is the same as…

Student: $^6/_8$

AV: See if you can come up with a formula for that. How do you get from $^3/_4$ to $^6/_8$?

Student: Multiply the top and bottom by 2?

AV: Good. What would happen if you multiplied the top and bottom by 3?

Student: You get $^9/_{12}$.

AV: Is that the same as $^3/_4$?

Student: Yes. Wait. I'm not sure.

AV: Draw a picture.

Student: Cut the circle into 12 parts?

AV: First cut it into four parts. Then cut each of those parts into three parts. <*The student does so.*> Now shade in…how many parts?

Student: 9?

AV: Good. <*The student shades in 9 parts.*> Is that the same as $^3/_4$?

Student: Yes. So you can multiply the top and bottom by any number you want?

AV: Exactly.

Student: And it's the same.

AV: Good. Do this. Give me five numbers that are the same as $^1/_2$.

Student: You mean do the same thing with multiplying?

AV: Yes.

Student: Like $^2/_4$?

AV: Yes.

Student: Or $3/6$?

AV: Yes.

Student: $4/8$, $5/10$, $6/12$, $7/14$.

AV: Good. What is $1/3 + 1/6$?

Student: Should I change the $1/3$ or the $1/6$?

AV: Up to you. < *The student tries out a few things, or thinks silently for a minute or two.* >

Student: $1/3$ is the same as $2/6$. So it is $2/6 + 1/6$, so it is $3/6$.

AV: Good. Do you know how to reduce that fraction?

Student: I've heard of it.

AV: What are you allowed to do to the top and bottom of a fraction?

Student: Multiply by the same number.

AV: Or?

Student: Add the same number?

AV: Try it. Start with $1/2$. Add something to the top and bottom.

Student: Like 2? So add 2 to the top and bottom. That is $3/4$.

AV: Is $1/2$ the same as $3/4$?

Student: No.

AV: Show me. Draw a picture of $1/2$ and a picture of $3/4$.

Student: With the circles and pie slices?

AV: Yes. < *The student draws the picture.* > Good.

Student: So you can't add the same number to the top and bottom. Can you divide?

AV: Maybe. Try $3/6$. What can you divide the top and bottom by?

Student: 3?

AV: OK, do it. Divide the top and bottom by 3. What do you get?

Student: $^1/_2$.

AV: Is $^1/_2$ the same as $^3/_6$?

Student: Yes.

AV: Show me a picture. <*The student draws two circles. The first is divided into six parts, and he shades in three of them. The second is divided into two parts, and he shades in one of them.*>

AV: Good. Now look at that question again. Do $^1/_3$ + $^1/_6$.

Student: Okay, $^1/_3$ + $^1/_6$ is the same as $^2/_6$ + $^1/_6$, which equals $^3/_6$. $^3/_6$ is the same as $^1/_2$, so the answer is $^1/_2$.

AV: Good. Now do $^1/_2$ + $^1/_6$.

Student: <*The student thinks for a few moments.*> $^1/_2$ is the same as $^3/_6$.

AV: Why?

Student: Because you can multiply the top and bottom by 3.

AV: Good.

Student: So you have $^3/_6$ + $^1/_6$, which is $^4/_6$.

AV: Good.

Student: $^4/_6$. Can I do something with that?

AV: Good question.

Student: Oh, I can divide the top and bottom by 2.

AV: Good.

Student: So I get $^2/_3$.

AV: Good. Now do $^4/_7$ + $^3/_{14}$.

Student: $^4/_7$ is the same as $^8/_{14}$. $^8/_{14}$ + $^3/_{14}$ is the same as $^{11}/_{14}$.

AV: Good. What about $^1/_2$ + $^1/_3$?

Student: Okay…<*The student pauses for a minute or two while he thinks.*> Wait, do I need to change both of the fractions?

AV: Maybe.

Student: Wait. Okay, $1/2$ is the same as $3/6$. $1/3$ is the same as $2/6$. So is it $5/6$?

AV: Good. In that case, what was the common denominator?

Student: 6.

AV: Good. Now do $2/7 + 1/5$.

Student: <*Thinks for a minute two*> Wait, is the common denominator 35?

AV: Apparently.

Student: So $2/7$ is the same as $10/35$. And $1/5$ is the same as $7/35$. So the answer is $17/35$.

AV:. Good. Can you reduce that?

Student: <*The student tries for several moments.*> No.

AV: Good. Now do $1/4 + 1/6$.

Student: So…wait. Is the common denominator 12 or 24?

AV: Whichever you think is easier.

Student: 12?

AV: Good. See how 12 is smaller? We call it the least common denominator, because it is the least.

Student: Oh yeah, I remember my teacher used to talk about that.

AV: Good. The point is, you can use any common denominator. You can use 12 or 24. But it is a bit easier to use 12.

Student: Ok. So 12. So $1/4$ is the same as…$3/12$. And $1/6$ is the same as $2/12$. So the answer is $5/12$.

AV: Good. Now lets do this: $1/a + 1/b$.

Student: What?

AV: Start by finding the common denominator.

Student: How?

AV: What is the common denominator when you do $1/2 + 1/3$?

Student: 6. So is the common denominator... a*b?

AV: Good. You can just write ab.

Student: How can I make the denominator ab?

AV: Good question. How can you make the denominator ab.?

Student: With $\frac{1}{a}$...do I multiply the top and bottom by b?

AV: See if it works.

Student: Okay. So the top is... is the top b?

AV: What is 1*b?

Student: b, right?

AV: Maybe. What is 1*5?

Student: 5. So yeah, it is b.

AV: Good.

Student: So $\frac{1}{a}$ is the same as $\frac{b}{ab}$?

AV: Good. And...?

Student: And $\frac{1}{b}$ is... $\frac{1}{b}$ is the same as $\frac{a}{ab}$?

AV: Good.

Student: So $\frac{1}{a} + \frac{1}{b}$ is the same as $\frac{b}{ab} + \frac{a}{ab}$... which is... wait, is it $\frac{ab}{ab}$?

AV: How's that?

Student: Because a+b is ab.

AV: It is? What does ab mean?

Student: Oh, ab is a*b, not a+b. So then what is the top?

AV: Good question.

Student: Then is it just... a+b?

AV: Good.

Student: so the answer is $\frac{a+b}{ab}$?

AV: Good. Now do $\frac{1}{x} + \frac{1}{m}$.

Student: So is the common denominator xm?

AV: Is it?

Student: Yeah.

AV: Continue.

Student: So $\frac{1}{x}$ is the same as $\frac{m}{xm}$, and $\frac{1}{m}$ is the same as $\frac{x}{xm}$. So the answer is $\frac{m+x}{xm}$.

AV: Good. Now do $\frac{1}{x} + \frac{1}{2x}$

Student: Is the common denominator... is it just 2x?

AV: Good question. Is it?

Student: Is $\frac{1}{x}$ the same as $\frac{2}{2x}$?

AV: How's that?

Student: Because I multiplied the top and bottom by 2.

AV: Are you allowed to do that?

Student: Yes.

AV: Good.

Student: Ok so $\frac{1}{x}$ is the same as $\frac{2}{2x}$. So it is $\frac{2}{2x} + \frac{1}{2x}$. So is it just $\frac{3}{2x}$.

AV: Good. How about $\frac{1}{x} + \frac{1}{2m}$.

Student: Is the common denominator... is it 2mx?

AV: Probably.

Student: So $\frac{1}{x}$ is the same as $\frac{2m}{2mx}$, and $\frac{1}{2m}$ is the same as $\frac{x}{2mx}$. So the answer is $\frac{2m+x}{2mx}$.

AV: Good. Now do $\frac{1}{x} + \frac{1}{(x+2)}$

Student: The common denominator... is the common denominator x+2?

AV: Explain.

Student: Because you can... wait, no it's not. You can't add the same thing to the top and bottom.

AV: What can you do?

Student: Multiply.

AV: Or?

Student: Divide. Wait, is the common denominator x*(x+2)?

AV: It does seem that way.

Student: So, for the first one multiply the top and bottom by (x+2), so you get $^{(x+2)}/_{x^*(x+2)}$. And the other is just $^x/_{(x+2)}$. So it is $^{x+2+x}/_{x^*(x+2)}$.

AV: Good simplify the top.

Student: Is it $x^2 + 2$?

AV: Is x+x the same as x^2

Student: Yeah.

AV: So 4+4 is 4^2?

Student: No. Wait. No 4*4 is 4^2, not 4+4.

AV: Good. What, then is x+x.

Student: Is it 2x?

AV: Is 4+4 the same as 2*4.

Student: < *The student pauses for several moments.* > Yeah.

AV: Good.

Student: Is the answer $^{2x+2}/_{x^*(x+2)}$

AV: Good. Also, you don't have to write x*(x+2). You can just write x(x+2).

Student: Okay. Yeah, I knew that before.

AV: Good. Now do $^2/_{(x+2)} + ^3/_{(x+1)}$

Student: So the common denominator is (x+2)(x+1). $^2/_{(x+2)}$ is the same as $^{2(x+1)}/_{(x+2)(x+1)}$. $^3/_{(x+1)}$ is the same as $^{3(x+2)}/_{(x+2)(x+1)}$. So the answer is $^{2(x+1) + 3(x+2)}/_{(x+2)(x+1)}$.

AV: Good. We'll talk about how to simplify the top in a few minutes. But first, we are going to do the actual problem. Do $^1/_x + ^1/_{(x+7)}$

Student: So the common denominator is x(x+7). So it is $^{(x+7)}/_{x(x+7)} + ^x/_{x(x+7)}$. So the answer is $^{x+7+x}/_{x(x+7)}$. So that is $^{2x+7}/_{x(x+7)}$.

That sure seems like a lot to go through to teach one problem, especially when you could give a one minute explanation like this:

1. Multiply the top and bottom of each fraction by the opposite denominator. Then add the numerators (the top parts) of each fraction, and write that sum over the common denominator.

Most students can memorize a set of steps like that for a quiz or test. The problem is that none of the underlying issues have been addressed. The student would still have no real understanding of fractions. He would not have developed the relevant visualization skills, or seen the connections between the easier numerical fractions and the more difficult algebraic fractions. He would memorize the formula, but not truly understand what he was dealing with. In fact, he would remember it by rote, the way you would remember "When you see a faquat, hit it with a potwu."

Using such a formula, he might even get an A on a quiz. But his exam scores would be much worse, and his SAT math scores would be horrendous. Standardized tests like the SAT are designed to determine a student's real level of understanding, and students who lack understanding of fundamental math concepts do extremely poorly.

When you use the Microchallenge method, the student's real math skills develop. Slowly but surely, he becomes smarter. With the above one minute explanation, the student develops no real skills; he merely learns a meaningless sequence of steps.

In the worst cases, teachers develop clever rules of thumb that allow students to do the problem without having to develop even the slightest understanding of any of the underlying concepts or methods. For example, for a problem like $1/x + 1/(x+7)$, the explanation might be like this:

> "When you are adding two fractions, and both have a '1' in the numerator, the answer will be this: On the top, add the two denominators. On the bottom, multiply the two denominators."

That formula is sort of clever, and it does work. But it does not lead to any real understanding or any kind of real cognitive development. Instead, it leads to temporary rote memorization and distancing. Again, the student memorizes this formula the way you might remember, "when you see a faquat, hit it with a potwu." It means nothing to him; it is something to remember temporarily and forget as soon as possible.

Students who learn these kinds of rules of thumb instead of real math do horrendously on exams and standardized tests. Additionally, they do not develop the cognitive skills, including spatial and logical reasoning skills, that a superior math education builds.

When using the Microchallenge method, never tell a student how to do the problem. If he does not know how to do something, give him an easier problem that has some of the same principles. If he cannot add $1/a + 1/b$, have him add $1/2 + 1/3$. But do not give direct explanations for anything. Make the student figure out every part of the process. This ensures that he develops a thorough understanding of the math concepts, and the true confidence that comes with that understanding.

Make sure that the student understands each concept before moving on. Give him two or three problems to make sure that he is incorporating each principle into his permanent understanding rather than temporarily memorizing it.

Finally, stay patient. Say you have been working for an hour, and the student almost has solved the problem. He just makes one small mistake. You will be tempted to just explain it so you can be done with it. Do not give in to that temptation. Stick to the Microchallenge method – use questions to guide him, but do not give a direct explanation. Help him develop his cognitive skills; do not give him a tempting way to avoid developing them.

As part of staying patient, accept the fact that you will often have to use the microchallenge method several times for the same type of problem. Each time it may become faster, but you may have to go through that kind of process five or ten times for each concept. However, as the student repeatedly goes through the process, he will gradually evolve into an excellent math student.

Even if you never excelled in trigonometry or calculus, you may be able to use the Microchallenge method for basic algebra. Many parents who feel that they are not good at math actually have the knowledge to use the Microchallenge method for many areas of math. Because most parents learned to do math without a calculator, their fundamentals are strong, even if their understanding of advanced math concepts is limited. Use the Microchallenge method as much as you can; when you cannot use it, switch to another method, such as the Asian system.

BUILDING THE MATHEMATICAL SCHEMA

As you teach your child math, you will notice that he learns some concepts quickly, but repeatedly forgets others. It may seem that some concepts just leak out of the child's mind.

Parents and teachers often find this situation intensely frustrating. Even the most patient teachers may feel their nerves begin to fray after the twentieth time a student forgets how to add fractions or forgets how to solve a quadratic equation.

When I discover that there is something specific that a student keeps forgetting, I have just the opposite reaction. From my perspective, I have just discovered a way to dramatically improve the child's mathematical abilities. I know that if I target that specific problem area, the child's abilities will improve in a fundamental and profound way.

We all have a complex system of processing mathematical information. I call this system the mathematical schema. The mathematical schema is highly resistant to change. Thus, new mathematical information is generally understood in a way that allows it to be easily incorporated into the current schema.

A child will find some information easy to incorporate into his schema, and find other concepts difficult to incorporate into his schema. In fact, some information will actually be impossible for the child to incorporate into his current mathematical schema. In other words, the new concept just does not fit into his current understanding of mathematics. Instead of incorporating this information into his schema, he will temporarily memorize it, and then forget it quickly.

For example, a child may be unable to think of $3x + 2$ as a single entity. In his mind, $3x + 2$ is two separate terms. Thus, he will be unable to understand how to do the following problem:

$f(x) = x^2 + 2x + 7$

Find f(3x+2).

The correct answer is

$$(3x+2)^2 + 2(3x+2) + 7.$$

The child will be able to memorize this in the short term, and may even gain some kind of temporary understanding. However, a day later he will have forgotten how to do the problem since he was never able to properly incorporate it into his mathematical schema.

For the child to learn a concept that does not fit with his mathematical schema, the schema itself must develop. This is great news; when the schema expands, the child's abilities improve at a fundamental level.

Clever explanations are not enough to cause the schema to expand permanently. The student must be repeatedly exposed to the problem type. Only when the mind realizes that it has no option other than to permanently expand the schema will the schema grow.

The method is simple. As soon as you find a problem type that the student forgets how to do, just give him that type of problem once a day. After a few weeks, he will be able to do the problem easily. At that point, his schema has been permanently expanded. He is now fundamentally smarter than he was before.

The Hydra is a great way to accelerate the process (in the Hydra, every time the child gets a problem wrong, he gets two additional similar problems. If he gets one of those wrong, he gets two more similar problems. Note that if he gets a problem wrong, he must redo the original problem in addition to the two new ones.) If the child has forgotten a particular concept, he might end up doing dozens, or even hundreds, of problems during a given Hydra drill. This gives the mind a powerful incentive to incorporate the new information as rapidly as possible.

14: THE REAL EQUATION

When a weak math student undergoes intensive training to become an excellent math student, more changes than just his math ability. Often, the student realizes that many of the other limitations he once thought he had do not really exist.

That, I believe, is why so many of my students start working out, eating healthy food, and reading on their own. Deep down, every child wants to believe that he can achieve excellence in all areas of life. But achieving that kind of excellence can seem so difficult that many write it off as impossible. Most of what I do is help students see the difference between what is extremely difficult and what is impossible. When a student learns that a goal which he once considered impossible is merely difficult, he is much more ready to pursue it with zeal.

In fact, at the core, that is what *The Equation for Excellence* is really about: drawing the distinction between the difficult and the impossible, and learning to achieve that which is difficult. If your child struggles with math, he will face a difficult road to excellence. At times, you will find it difficult to teach him. But it will not be impossible. And as you guide him through this journey, you may find yourself realizing that some of the limits that you have set for yourself are just as nonexistent as your child's.

ABOUT THE AUTHOR

One of America's most important eduational innovators, Arvin Vohra has transformed the lives of hundreds of students with his rigorous and inspiring approach to math. His company, Arvin Vohra Education (AVE), provides individuals and institutions with groundbreaking educational solutions, including the Arvin Vohra Accelerated Math Curriculum, the Rapid Analytical Reading Method, and the Integrated SAT Math Curriclum.

Arvin Vohra's passion for educational innovation began at an early age. He learned algebra in a few weeks to advance a level in math, and even attended college courses as an eighth grader. In high school, he received a score of 5 on ten AP exams to become an AP National Scholar. For six of those exams, he did not take the corresponding AP class, and instead relied on intensive independent study. He was also a National Merit Finalist, with the highest SAT and PSAT scores in his graduating class. During this time, he worked actively as a tutor for younger students.

At Brown University, his passion for educational innovation continued to grow. He worked as a tutor and teacher as part of the Wheeler School's Aerie Program, where he designed accelerated and nontraditional curricula for students ranging from second to eighth grade. Later, he worked as a consultant for the Hamilton Institute for Learning Differences. After graduating from Brown University with a B.Sc. in Mathematics, he received a perfect score on both the GRE and the GMAT. He currently resides in Bethesda, MD.